Praise for *The Expecta*



"As David Robson makes plain in this comp about the world can profoundly shape h[illegible] Based in science and packed with smart advice, *The Expectation Effect* will expand your mind—and maybe even extend your life."

—Daniel Pink, *New York Times* bestselling author of *When*, *Drive*, and *To Sell Is Human*

"This is an utterly riveting and transformative book. You can't afford not to read it!" —Nigella Lawson

"Visionary, original, and exciting, like all the best science writing."

—Will Storr, author of *The Science of Storytelling*

"A fascinating exploration of the ways in which our brains can sabotage or save us in our most challenging moments." —*Salon*

"[*The Expectation Effect*] made me realize that I'm leveraging far fewer of my brain's features than I could be. . . . What scientists are learning about the connections between the human brain and performance is incredible. And David Robson manages to take this research and organize it into a compelling playbook for a better life."

—Gayle Allen, *Curious Minds at Work* podcast

"Whether you're looking to make major or minor changes in your life, this book will help you leave the starting gate with positive expectations of success." —*New York Journal of Books*

"Citing examples across the spectrum of human behavior, Robson shares compelling studies and anecdotal evidence to suggest that what people think will happen often turns into a self-fulfilling prophecy. . . . In *The Expectation Effect*, he more than meets his goals of equipping readers with a critical understanding of how expectations shape lives and providing practical tools for reframing and adjusting mindsets toward positivity and well-being." —*Shelf Awareness*

"The human brain, according to this absorbing book, has a mind of its own. . . . Cutting-edge research and effective storytelling create an insightful book on an ever-changing field." —*Kirkus Reviews*

"A fascinating, optimistic, and empowering book. David Robson uses science and stories to show the extent to which we are shaped by our beliefs, and how the predictive power of our brains influences pain, diet, sleep, exercise, and intelligence." —Dr. Monty Lyman, author of *The Remarkable Life of the Skin* and *The Painful Truth*

"Authoritative, measured, practical, and encouraging, it will change your attitude to life's challenges."

—Dr. Christian Jarrett, author of *Be Who You Want*

"An intriguing account of the role of expectation (and perception in general) in a wide panorama of experience. Beautifully written, science-based, and a gripping read. I loved it!"

—Dr. Mithu Storoni, author of *Stress-Proof*

"As a counter to the many 'pseudoscientific' self-help books about the power of expectations (most notably, *The Secret*), Robson investigates the 'expectation effect' through peer-reviewed experiments and studies. . . . A fine place to start for readers interested in the power of the mind." —*Publishers Weekly*

"The consistently brilliant David Robson is back with science, stories, and practical tips for anyone who wants to make better use of the remarkable body-mind connection." —Kerry Daynes, bestselling author of *The Dark Side of the Mind* and *What Lies Buried*

"This book is entertaining, eye-opening, extremely useful, and, best of all, evidence-based." —Claudia Hammond, author of *The Art of Rest*

"*The Expectation Effect* is a must-read book that shows how we can all use our brainpower to improve our lives. Whether it's medicine, anxiety, exercise, or diet, David Robson seamlessly and engagingly explores how what we expect to happen has a much larger role than we might have previously believed—and we can tap into that power to improve our own well-being. . . . If you want to learn how to reframe your mindset, to feel healthier and happier in the process, this book is for you." —Melissa Hogenboom, author of *The Motherhood Complex*

"How on Earth does a person pack in so much research and yet still make it fascinating and fun to read? Through the perfect mix of stories, stats, and science, David Robson makes a very persuasive case that what we expect has a significant material impact on our lives—and how we can harness this to live longer, happier, healthier, more successful lives." —James Wallman, author of *Stuffocation* and *Time and How to Spend It*

"Mind-changing science . . . One of Robson's many strengths as a chronicler of science is to take what might seem familiar and show . . . just how much deeper the rabbit hole goes. . . . The book abounds in compelling anecdotes. . . . Robson's central point is that the expectation effect isn't an amusing psychological quirk, but a fundamental aspect of our interactions with reality. . . . The result of marinating for a while in this outlook is surprisingly transformative."

—Oliver Burkeman, *The Guardian*

ALSO BY DAVID ROBSON

*The Intelligence Trap*

# THE EXPECTATION EFFECT

# THE EXPECTATION EFFECT

*How Your Mindset Can Change Your World*

DAVID ROBSON

A HOLT PAPERBACK
HENRY HOLT AND COMPANY
NEW YORK

Holt Paperbacks
Henry Holt and Company
*Publishers since 1866*
120 Broadway
New York, New York 10271
www.henryholt.com

A Holt Paperback® and H® are registered trademarks of
Macmillan Publishing Group, LLC.



Distributed in Canada by Raincoast Book Distribution Limited

The Library of Congress has cataloged the hardcover edition as follows:

Names: Robson, David (Science journalist), author.
Title: The expectation effect : how your mindset can change your world / David Robson.
Description: First US edition. | New York : Henry Holt and Company, [2022] | Includes bibliographical references and index.
Identifiers: LCCN 2021047473 (print) | LCCN 2021047474 (ebook) | ISBN 9781250827630 (hardcover) | ISBN 9781250827647 (ebook)
Subjects: LCSH: Expectation (Psychology) | Attitude (Psychology) | Self-actualization (Psychology)
Classification: LCC BF323.E8 R63 2022 (print) | LCC BF323.E8 (ebook) | DDC 158.1—dc23
LC record available at https://lccn.loc.gov/2021047473
LC ebook record available at https://lccn.loc.gov/2021047474

ISBN: 9781250871091 (trade paperback)

Our books may be purchased in bulk for promotional, educational, or business use. Please contact your local bookseller or the Macmillan Corporate and Premium Sales Department at (800) 221-7945, extension 5442, or by email at MacmillanSpecialMarkets@macmillan.com.

Originally published in hardcover in 2022 by Henry Holt and Company

First Holt Paperbacks Edition 2023

Designed by Gabriel Guma

Printed in the United States of America

D 10 9 8 7 6 5

# CONTENTS

## LIST OF ILLUSTRATIONS AND GRAPHS

# THE EXPECTATION EFFECT

# INTRODUCTION

> The mind is its own place and in it self
> Can make a Heaven of Hell, a Hell of Heaven.
>
> –John Milton, *Paradise Lost*

Our expectations are like the air we breathe—they accompany us everywhere, yet we are rarely conscious of their presence. You might assume that your body is resilient, or that it is prone to sickness. You might think you are naturally lean and sporty, or that you are predisposed to gaining weight. You might believe that the stresses in your life are harming your health, and that a night of poor sleep will render you a walking zombie the next day.

These assumptions may appear to be inescapable, objective truths. But in this book I will show you how those beliefs, *in themselves*, shape your health and well-being in profound ways, and that learning to reset our expectations about these issues can have truly remarkable effects on our health, happiness, and productivity.

Don't believe me? Then consider one attention-grabbing study from Harvard University. The participants were hotel cleaners, whose work is often physically intense yet feels very different from the exercise you might perform at the gym. To change the cleaners' perceptions of their own fitness, the researchers explained the amount of energy that was exerted by vacuuming the floor, changing beds, or moving furniture—which, over the course of a week, easily amounts to the level of exercise recommended for good health. One month later, the researchers found that the cleaners' fitness had noticeably improved, with significant changes in their weight and blood pressure.[1] Quite amazingly, the shift in their beliefs about their bodies, and their new expectations of their work, had brought about real physiological benefits—without any change in lifestyle.

We will discover how "expectation effects" like this can also influence our susceptibility to illness, our ability to maintain a stable body weight, and the short- and long-term consequences of stress and insomnia. As the following story shows, the power of expectation is so strong that it can even determine how long you live.

Starting in the late 1970s, the US Centers for Disease Control began to receive reports that a worrying number of recent Laotian immigrants were dying in their sleep. They were almost all male, aged between their midtwenties and midforties, and most were from the persecuted Hmong ethnic group who had fled Laos after the rise to power of the Pathet Lao. For their loved ones, the only warning was the sound of them struggling for breath and, occasionally, a gasp, a moan, or a cry. By the time help arrived, however, they were already dead.

Try as they might, epidemiologists could find no good medical explanation for these "Sudden Unexpected Nocturnal Death Syndrome" cases. Autopsies showed no evidence of poisoning; nor was there anything particularly unusual about their diet or their mental health. At its peak, however, the mortality rate was so high among young Hmong men that SUNDS accounted for more lives lost than all the other top five causes of death combined. Why were so many seemingly healthy adults passing away in their sleep?

Investigations by the medical anthropologist Shelley Adler would eventually solve the mystery. According to Hmong traditional folklore, an evil demon called the "dab tsog" roamed the world at night. When it had found its victim, it would lie on the body, paralyzing the victim and smothering their mouth until they could no longer breathe.

Back in the mountains of Laos, the Hmong could ask a shaman to build a protective necklace, or they could sacrifice animals to appease their ancestors, who would fend off the dab tsog. But now these men were in the United States—there were no shamans, and they were no longer able to perform their ritual sacrifices to appease their ancestors, meaning they had no more protection from the dab tsog. Many had converted to Christianity so that they could better integrate into American culture, neglecting their old rituals altogether.

Guilt at having abandoned their traditions was itself a source of chronic stress that could have harmed their overall health. But it was at night that the fears of the dab tsog became a reality, with disturbing nightmares that resulted in the experience of sleep paralysis, in which the mind becomes conscious, as if you were fully awake, but the body is unable to move. Sleep paralysis is not in itself dangerous—it affects around 8 percent of people.[2] For the Hmong immigrants, however, it seemed like the dab tsog had come to wreak revenge. The result, Adler concluded, was a panic so strong it could exacerbate a heart arrhythmia, leading to cardiac arrest.[3] And as the deaths mounted, the Hmong men only became more scared, creating a kind of hysteria among the population that may have caused even more deaths. The explanation is now accepted by many scientists.[4]

Newspaper reports at the time described the "cultural primitivity" of these people, who were "frozen in time" and "ruled by superstition and myth." But scientists now argue that we are all susceptible to beliefs that are just as potent as the dab tsog. You may not believe in demons, but thoughts about fitness and expectations about long-term health may have real consequences for your longevity, including the risk of heart disease. This is the enormous power of the expectation effect. It is only once we recognize its influence that we can begin to use it to our advantage to ensure a longer, healthier, happier life.

These provocative claims may sound dangerously close to the content of many New Age self-help books, such as Rhonda Byrne's 35-million-copy best seller *The Secret*. Byrne promoted the "law of attraction"—the idea that, for example, visualizing yourself rich will bring more money into your life. Such ideas are pure pseudoscience, whereas the findings in this book are based on robust experiments published in peer-reviewed journals, and they can be explained by well-accepted psychological and physiological mechanisms, such as the actions of the nervous and immunological systems. We will learn how our beliefs can influence many important life outcomes without any appeal to the paranormal.

You may also wonder how the content of our thoughts could have any meaningful influence in the chaos of the world today. I wrote much of this book in the midst of the Covid-19 pandemic, when many of us were grieving for loved ones and fearing for our livelihoods. We have also faced huge political uncertainty and unrest, and many continue to wrestle with enormous structural inequalities. Our own expectations and beliefs may seem to hold little power in the face of all these barriers.

It would be foolish to argue that "positive thinking" could eliminate all this unhappiness and anxiety—and I would be the last person to make that claim. (Scientific research continues to show that simply denying the difficulties of a situation will only lead to worse outcomes.) As we shall soon see, however, there are many ways that our beliefs about our own capacities can influence how we cope with challenges, and can determine the toll they take on our physical and mental health. While many of today's crises are beyond our control, our responses to difficult situations are often the product of our expectations—and understanding this allows us to increase our resilience and to react in the most constructive way to the problems we face.

Crucially—and this is something that I will emphasize throughout the book—the expectation effects described in these chapters concern *specific* beliefs rather than a general optimism or pessimism. Armed with scientific knowledge about the ways your expectations are shaping your life, you can learn to reframe and reappraise your thinking without any self-deception, and you don't need to turn into a cheery Pollyanna to benefit.

My own understanding of the enormous power of expectations came seven years ago during a period of turmoil in my own life.

Like many people, I had previously suffered from depression and anxiety, but for most of my life I had managed to weather the waves of unhappiness until they passed. Then, after a period of intense stress, the troughs in my mood started to get deeper, and longer, to the point that they were no longer bearable.

Recognizing those symptoms, I went to my family doctor, who prescribed me a course of antidepressants and offered some of the usual warnings about the known side effects, including migraine. Sure enough, my mood seemed to stabilize, but on those first few days I also experienced splitting headaches that felt like an ice pick had penetrated my skull. The pain was so intense that I was sure that something terrible was happening to my brain. How could this agony not be some kind of warning?

It just so happened, however, that I had also started writing a popular science article about the placebo effect (from the Latin, "I shall please"). As is now well known, inert sugar pills can often reduce symptoms and accelerate recovery through the patient's mere expectation that they will heal the body, and this coincides with physiological changes to blood circulation, hormone balance, and immunological response.

While working on my article, I discovered that not only do many people taking placebo pills experience the benefits of the drug they believe they are taking; they also report its side effects, too, from nausea, headaches, and fainting to sometimes dangerous drops in blood pressure. And the more people are told about those side effects, the more likely they are to report them. These are known as nocebo effects (from the Latin, "I will harm"), and like the placebo responses, these symptoms are not "imagined" but are the result of measurable physiological changes—including significant shifts in our hormones and neurotransmitters.

For many antidepressants, the vast majority of side effects can be explained by the nocebo response rather than an inevitable reaction. In other words, the terrible pain I had been feeling while taking medication was perfectly real—but the product of my mind's expectation rather than the actual chemical effects of the drug. With this knowledge, the pain soon vanished. After a few more months of taking the antidepressant (side-effect–free), my depression and anxiety had lifted. Knowing that many of the symptoms of withdrawal may arise from the nocebo effect undoubtedly helped me to eventually wean myself off the medication, too.

Ever since, I have closely followed the research on the mind's capacity to shape our health and well-being and our physical and mental capacities. And it is now becoming apparent that the placebo and nocebo responses to drugs are just two examples of the ways that beliefs can become self-fulfilling prophecies, changing our lives for better or worse. In the scientific literature, these phenomena are variously called "expectation effects," "expectancy effects," "Oedipus effects" (after the self-fulfilling prophecy in Sophocles's famous play), and "meaning responses." For simplicity, I use the first—"expectation effects"—to describe all the scientific phenomena that underlie the real-world consequences of our beliefs.

The study of hotel cleaners is just one example of this cutting-edge research, but there are many other fascinating findings. So-called "complaining good sleepers"—people who vastly overestimate how much time they spend awake and restless each night—are much more likely to suffer greater fatigue and poor concentration during the day, while "non-complaining bad sleepers" seem to escape the ill effects of insomnia. For the purposes of our next-day performance, we slept as well as we think we did.

Beliefs about the consequences of anxiety, meanwhile, can change someone's physiological responses to stress, affecting both short-term performance and the long-term toll on mental and physical health. Positive and negative self-fulfilling prophecies can also determine memory capacity, concentration, and fatigue during hard mental tasks, and creativity in problem solving. Even someone's intelligence—long considered to be an immutable trait—may climb or fall according to their expectations.

These findings are causing some scientists to question the fundamental limits of the brain, suggesting that we may all have untapped mental reserves that we can free if we develop the right mindset. And that has immediate implications for work and education, and the ways that we cope under new pressures.

The most mind-blowing results concern the aging process. People with a more positive attitude to their later years are less likely to develop hearing loss, frailty, and illness—and even Alzheimer's disease—than people who associate aging with senility and disability. In a very real sense, we are as young as we feel inside.

As the Harvard study of the hotel cleaners shows, our expectations are not set in stone. Once we acknowledge the power that our expectations hold over our lives, the research offers some straightforward psychological techniques that we can apply to boost our physical and mental health and unleash our full intellectual potential. In the words of one of the most influential researchers in this field, Alia Crum at Stanford University: "Our minds aren't passive observers simply perceiving reality as it is; our minds actually change reality. In other words, the reality we will experience tomorrow is in part a product of the mindsets we hold today."[5]

So how do the body, brain, and culture interact so potently to produce these self-fulfilling prophecies? What are the beliefs and expectations that rule our physical and mental well-being? And how can we use these fascinating findings to our own benefit? These are the central questions this book sets out to answer.

We will begin this journey with a revolutionary new theory, which sees the brain as a "prediction machine" that constantly simulates the future. This theory can explain how conscious and unconscious expectations can powerfully influence our perceptions of reality—from the strange hallucinations of Arctic explorers to our experience of pain and illness. Importantly, this prediction machine can also alter our body's physiology—leading us to explore the power of belief in medicine, including an extraordinary psychological intervention that can accelerate your recovery from surgery. We will discover the ways that expectations can be transmitted between individuals through social contagion and can explain the psychosomatic origins of many recent health crises, including the perplexing rise in food allergies—and the ways to avoid falling victim to these expectation effects yourself.

We will then move beyond medicine to explore the power of expectation in everyday health and well-being. We will see how food labeling can change the way your body processes nutrients, with a direct impact on your waistline; how to use your mind to take the pain out of exercise and improve your athletic performance, without performance-enhancing drugs; and how to change your physical and mental responses to stress. We shall understand how prevailing cultural beliefs in countries such as India produce much better concentration and willpower. We will also learn the secrets of "super-agers" from the world's oldest acrobatic salsa dancer, and the powerful potential of belief to slow the ravages of time—right down to the aging of our individual cells. Finally, we will return to the Hmong and discover how their story can help us create our own self-fulfilling prophecies.

At the end of each chapter, you will also find summaries of the techniques to employ expectation effects to your advantage. These will vary in their details—but in general they will work better with repetition and practice. I encourage you to approach them with an open mind—testing the principles in comfortable situations, with the aim of building on any small gains. While it may be tempting to skip ahead to the practical news you can use, these expectation effects tend to be more potent if you understand the science behind their success. The more deeply you process the material, the greater the benefits—so it may also be useful to write down the specific ways you hope to apply it in your life. You may even like to share your results on social media, through the #expectationeffect hashtag, or upload them to the website www.expectationeffect.com, which I shall be updating regularly; there is some research suggesting that sharing an expectation effect with others—and hearing about their experiences in return—can increase its power.

Let me be absolutely clear. Your mind alone cannot perform miracles—you cannot simply imagine piles of money and think yourself rich, or cure yourself of a terminal illness through positive visualizations. But your expectations and beliefs can influence—indeed are already influencing—your life in many other surprising and powerful ways, and if you want to learn how to turn them to your advantage, read on. You may be astonished by your potential for personal change.

# 1

# THE PREDICTION MACHINE

## How your beliefs shape your reality

It was just a few nights before Christmas, and the drones seemed to be everywhere and nowhere at the same time.

The drama began at nine p.m. on December 19, 2018, when a security officer at London Gatwick Airport reported two unmanned aerial vehicles—one flying around the perimeter fence, another inside the complex. The runway was soon closed for fear of an impending terrorist attack. It was only nineteen months after the Islamist bombing at the Manchester Arena, after all, and there had been reports that members of ISIS were planning to carry explosives on commercial drones.

The chaos escalated over the following thirty hours as dozens of further sightings kept the airport in lockdown. Try as they might, however, the security officers and police just couldn't locate the drones, which seemed to disappear as soon as they were sighted. Even more astonishingly, their operators appeared to have found a way to avoid the military's track-and-disable system, which was unable to detect any unusual activity in the area, despite a total of 170 reported sightings. The news soon spread to the international media, which warned that similar attacks might occur in other countries.

By six a.m. on December 21, the threat finally seemed to have passed, and the airport reopened for business. Whoever was behind the attack—be it a terrorist or a joker—had achieved their aim of chaos, disrupting the travel of 140,000 passengers with the cancelation of more than a thousand flights. Despite offering a substantial reward, the British police have been completely unable to find a culprit, and there is not a single photo offering evidence of an attack—leading some (including members of the police) to question whether there were ever any drones at all.[1] Even if there was, at one point, a drone near the airport, it's clear that the vast majority of the sightings were false, and the ensuing chaos was almost certainly unnecessary.

With so many independent reports from dozens of sources, we can easily rule out the possibility that this was some kind of lie or conspiracy. Instead, the event demonstrates the power of expectation to change our perception, and—occasionally—to create a vision of something that is entirely false.

According to an increasing number of neuroscientists, the brain is a "prediction machine" that constructs an elaborate simulation of the world, based as much on its expectations and previous experiences as on the raw data hitting the senses. For most people, most of the time, these simulations coincide with objective reality, but they can sometimes stray far from what is actually in the physical world.[2]

Knowledge of the prediction machine, and its workings, can help us to understand everything from ghost sightings to disastrously bad calls by sports referees—and the mysterious appearance of nonexistent drones in the winter sky. It can help us to understand why the name we call a beer changes its taste; and it shows how to someone with a phobia the world looks much more terrifying than it really is. This grand new unifying theory of the brain also sets the stage for all the expectation effects that we'll examine in this book.

## THE ART OF SEEING

The seeds of this extraordinary conception of the brain were sown in the mid-nineteenth century by the German polymath Hermann von

Helmholtz. Studying the anatomy of the eyeball, he realized the patterns of light hitting the retina would be too confusing to enable us to recognize what is around us. The 3D world—with objects at various distances and odd angles—has been flattened onto two two-dimensional disks, resulting in obscured and overlapping contours that would be difficult to interpret. And even the same object may reflect very different colors depending on the light source. If you are reading this physical book indoors at dusk, for example, the page will be reflecting less light than a dark gray page in direct sunlight—yet in both cases, the page looks distinctly white.

Helmholtz suggested that the brain draws on past experiences to tidy up the visual mess and to come up with the best possible interpretation of what it receives, through a process he called "unconscious inference." We may think we are seeing the world unfiltered, but vision is really forged in the "dark background" of the mind, he proposed, based on what it assumes is most likely to be in front of you.[3]

Helmholtz's theories of optics influenced postimpressionist artists like Georges Seurat,[4] but it was only in the 1990s that the idea really started to take off in neuroscience—with signs that the brain's predictions influence every stage of visual processing.[5]

Before you walk into a room, your brain has already built many simulations of what might be there, which it then compares with what it actually encounters. At some points, the predictions may need retuning to better fit the data from the retina; at others, the brain's confidence in its predictions may be so strong that it chooses to discount some signals while accentuating others. Over numerous repetitions of this process, the brain arrives at a "best guess" of the scene. As Moshe Bar, a neuroscientist at Bar-Ilan University in Israel who has led much of this work, puts it: "We see what we predict, rather than what's out there."

A wealth of evidence now supports this hypothesis, right down to the brain's anatomy. If you look at the wiring of the visual cortex at the back of the head, you find that the nerves bringing electrical signals from the retina are vastly outnumbered by the neural connections feeding in predictions from other regions of the brain.[6] In terms of the data it provides, the eye is a relatively small (but admittedly essential) element of your vision, while the rest of what you see is created "in the dark" within your skull.

By measuring the brain's electrical activity, neuroscientists like Bar can watch the effects of our predictions in real time. He has observed, for instance, the passing of signals from frontal regions of the brain—which are involved in the formation of expectations—back into the visual cortex at the earliest stages of visual processing, long before the image pops into our consciousness.[7]

There are lots of good reasons why we might have evolved to see the world in this way. For one thing, the use of predictions to guide vision helps the brain to cut down the amount of sensory information it processes, so that it can focus on the most important details—the things that are most surprising, and which do not fit its current simulations.

As Helmholtz originally noted, the brain's reliance on prediction can also help us to deal with incredible ambiguity.[8] If you look at the

image on page 14—a real, albeit poor-quality, bleached photograph—you will probably struggle to identify anything recognizable.

If I tell you to look for a cow, however—facing you, with its large head toward the left of the image—you may find that something somehow "clicks," and the image suddenly makes a lot more sense. If so, you've just experienced your brain's predictive processing retuning its mental models to make use of additional knowledge, transforming the picture into something meaningful.

Or what do you see when you look at the image below? (Try for at least ten seconds before going on.)

If you're like me, you will initially find it extremely hard to make out anything specific. But if you see the original image (on page 33) it suddenly becomes a lot clearer: that's your brain's updated predictions making sense of the mess.[9] Once you've seen the original, it's almost impossible to believe that you were ever confused by the unclear image—and the effect of those updated predictions is enduring. Even if you return to this page in a year's time, you'll be much more likely to see the dog than when you first saw the incomprehensible splotches of black and white.

The brain will draw on any contextual information it can to refine its predictions—with immediate consequences for what we see. (If you'd seen the picture in a pet store or a veterinary surgeon's office, you might have been far more likely to have seen the dog at first glance.) Even the day of the year can determine how your brain processes ambiguous sights. A pair of Swiss scientists, for example, stood at the main entrance of Zoo Zürich and asked participants what they saw when they looked at a version of a famously ambiguous visual illusion (reproduced below).

In October, around 90 percent of the zoo visitors reported seeing a bird looking to the left. At Easter, however, that dropped to 20 percent, while the vast majority saw it as a rabbit looking to the right. Of children under ten, for whom the Easter Bunny may be an especially important figure, nearly 100 percent saw a rabbit on the holiday weekend. The prediction machine had weighed up which potential interpretation of the ambiguous picture was most relevant, and the season managed to tip the balance—with a tangible effect on people's conscious visual experience.[10]

Psychologists sometimes describe our expectations as having a "top-down" influence, as opposed to the raw data flowing "bottom-up"

from the body. We now know that top-down influences are not limited to vision but govern all kinds of sensory perception. And it is an incredibly effective way of experiencing the world. Suppose you are driving on a misty day: if you are familiar with the route, your previous experiences will help your brain to make out the sight of a road sign or another car, so that you avoid having an accident. Or imagine you are trying to work out the meaning of someone's words on a crackly telephone line. This will be much easier if you are already familiar with the accent and cadences of the speaker's voice, thanks to the prediction machine.

By predicting the effects of our movements, the brain can damp down the feeling of touch when one part of our body makes contact with another, so that we don't jump out of our skin whenever one of our legs brushes against the other, or our arm touches our side. (It is also for this very reason that we can't tickle ourselves.) Errors in people's internal simulations might also explain why amputees still often feel pain in their missing limbs—the brain hasn't fully updated its map of the body and erroneously predicts that the arm or limb is in great distress.

There will inevitably be some small errors in each of the brain's simulations of the world around us—a mistaken object or a misheard sentence that is soon corrected. Occasionally, however, those simulations can go completely awry, with heightened expectations evoking vivid illusions of things that do not exist in the real world—such as drones flying over the UK's second-biggest airport.

In one brilliant demonstration of this possibility, participants were asked to watch a screen of random gray dots (like the "snow" on an untuned analog TV). With a suitable suggestion, they could be primed to see faces in 34 percent of trials, even though there was nothing there apart from random visual noise. The expectation that a face would appear led the brain to sharpen certain patterns of pixels in the sea of gray, leading people to hallucinate a meaningful image with astonishing frequency. What's more, brain scans showed the brain forming these hallucinations in real time, with the participants demonstrating heightened neural activity in the regions normally associated with face perception.[11] Clearly, seeing isn't believing—*believing is seeing*.

Believing is also hearing. In another study, Dutch researchers played some white noise to students under the belief that they could hear a very faint rendition of "White Christmas" by Bing Crosby. Despite the fact that objectively there was not a hint of music, nearly one-third of participants reported that they could really hear the song.[12] The implanted belief about what they were about to hear led the students' brains to process the white noise differently, accentuating some elements while muting others, until they hallucinated the sound of Crosby singing. Interestingly, a follow-up study found that auditory hallucinations of this kind are more common when we feel stressed and have consumed caffeine, which is thought to be a mildly hallucinogenic substance and may lead the brain to place more confidence in its predictions.

If you put yourself in the shoes of the officers at Gatwick, it's easy to imagine how fears of an impending terrorist attack could conjure up the image of a drone in the gray blanket of the winter sky, where there might be many ambiguous figures—birds or helicopters, for example—that the prediction machine is liable to misinterpret. And the more sightings that were reported, the more people would have expected to see further drones. Had scientists been able to peer into their brains, it is likely that they would have observed exactly the same brain activity that would be present in someone looking at an actual drone.[13]

Momentary hallucinations of this kind can result from the prediction machine's errors in countless other situations. Strange visions are apparently common among polar explorers, for example, as the unchanging blankness of the landscape—the "white darkness," as some describe it—plays havoc with the prediction machine's simulations.

One of the most memorable examples of this phenomenon concerns Roald Amundsen's expedition to Antarctica. On December 13, 1911,

Amundsen's team was within spitting distance of the South Pole, and dreaded the thought that Robert Falcon Scott's competing expedition might beat them to their goal. As they made up camp, one of Amundsen's group, Sverre Helge Hassel, called out that he had spotted people moving in the distance. Soon the whole team could see them. When the explorers ran forward, however, they soon discovered that it was a pile of their own dogs' turds lying on the snow. The explorers' minds had transformed a pile of feces into the thing they feared.[14]

Many supposedly paranormal experiences may arise through a similar process. When a fire broke out at Notre Dame Cathedral in Paris on April 15, 2019, for example, a number of eyewitnesses reported seeing the form of Jesus in the flames.[15] Some assumed it was a sign of God's disapproval at the turn of events; others, that He was trying to offer comfort to those affected by the damage. But scientists would argue that it was the observers' underlying beliefs that led their brains to construct something meaningful from ambiguous patterns of light. Whenever someone claims to have seen a ghost, or to have heard the voices of the undead in the static of an untuned radio, or to have seen an image of Elvis in the clouds, the overreactive prediction machine may be to blame. These phenomena are natural consequences of the way the brain normally makes sense of the world, although they are, of course, much more likely if you already hold religious or paranormal beliefs.

Athletes and referees would do well to remember the role of the prediction machine during sports controversies. When a tennis player and an umpire quarrel over a point, it reflects a serious difference in the perceptual experience: one "saw" the ball inside the lines and the other "saw" it outside. Neither party is being stupid or dishonest—their minds have just constructed different simulations of the world around them, causing them to have radically different experiences of the event. For each person, the perception could have seemed just as "real" as the greenness of the grass or the blueness of the sky. A confident player in particular might be primed to see the ball landing in their favor, and without any conscious intention to deceive that could sway their perception—a phenomenon that psychologists call "wishful seeing."[16]

At the time of the airport "attack," the police at Gatwick were keen to stress the credibility of their eyewitnesses, but the theory of the brain as a prediction machine suggests that there may be no such thing as a fully objective observer. As the neuroscientist Anil Seth puts it: "We don't just passively perceive the world, we actively generate it. The world we experience comes as much, if not more, from the inside out as from the outside in."[17] Our brain's expectations are intricately woven into everything we experience.

The philosophical implications of this inherent subjectivity are profound enough. But as we shall soon find out, the theory of the brain as a prediction machine also has immense consequences for our well-being—providing insights that go far beyond the emergence of uncanny visual illusions. And to see how, we need to meet a remarkable patient.

## "I WAS BLIND BUT NOW I SEE"

A young woman I'll call Sara was in her late teens when she woke up almost completely blind. Her vision had been deteriorating for six months; now she saw just a faint glow around certain light sources—everything else was darkness. Ophthalmologists could find nothing wrong with her eyes, though that knowledge did absolutely nothing to help her day-to-day well-being, as she carefully counted each step and felt her way around furniture to navigate her home.

After numerous examinations, Sara was diagnosed with a "functional neurological disorder" (FND), a term describing a serious problem with the brain and nervous system's workings without any evidence of anatomical damage. Other examples include deafness, a loss of sensation or movement in the limbs, or an inability to feel pain—all in otherwise physiologically healthy individuals. And these conditions are not as rare as you might assume: despite the relatively low public awareness, FNDs are in fact the second most common reason for someone to be referred to a neurologist (after migraines and headaches).[18] Sigmund Freud assumed that these symptoms were the consequence of repressed stress or trauma. Today many neurologists believe that FNDs such as Sara's may be a direct result of errors in the brain's predictions, which somehow dampen the normal processing of sensory signals to the point that they are no longer experienced. In Sara's case, her brain was effectively pulling shutters over her eyes.

She initially balked at the suggestion that her condition had "psychogenic" origins; it seemed a bizarre diagnosis when she had never previously experienced any psychiatric disorders, and when she was even dealing with her newly developed blindness with remarkable resilience. But she was eventually referred to Jon Stone, a neurologist at the University of Edinburgh who specializes in FNDs. During their initial conversations, he discovered that before losing her sight Sara had been experiencing chronic migraines that seemed to be triggered by light. This had led her to spend more and more time in a darkened room, until one morning she woke up without any vision at all.

Stone proposed that with her increasing "photophobia" (fear of light) and the constant seeking of darkness Sara's brain had somehow become stuck on the idea that it couldn't see anything. And while that erroneous expectation may have arisen unconsciously, Stone hoped that it would be possible to correct the error through continued encouragement and discussion. To do so, he would point out whenever Sara made eye contact with him or copied certain gestures—evidence that, unconsciously, her brain was still able to process some visual information—and he encouraged her family to do the same at home.

As further encouragement, Stone also employed a noninvasive form of brain stimulation in which an electromagnetic coil placed on the scalp excites neurons underneath the skull. Amplifying the electrical

With time—and a change of my apartment's locks—I stopped hearing these phantom break-ins. But there is now strong evidence that many enduring anxieties and phobias are accompanied—and perhaps partly caused—by permanently distorted perceptions of potential dangers in the environment. People with a fear of heights, for instance, were asked to look over a twenty-six-foot-high balcony and guess the distance to the ground. On average, their estimates were about five feet higher than those of people without the sense of fear.[21] Similarly, people with arachnophobia consistently see spiders as being much larger and faster than they actually are—and the greater their fears, the more pronounced the illusion.[22] When it is lurking on the wall next to you, a regular house spider can begin to look a lot like a menacing tarantula.

Warped perceptions, resulting from biases in the brain's predictions, can also contribute to our social anxieties. When people feel shy, sad, or nervous, they tend to see photos of neutral faces as looking more hostile compared with people who are in a calmer state of mind.[23] To make matters worse, the (conscious or unconscious) expectation of rejection leads them to linger on the potentially unfriendly faces for longer, while ignoring any friendly smiles. In one memorable experiment, psychologists tracked a group of university students' eye movements as they watched videos of adolescents during their school breaks. They found that someone's social success powerfully altered their experiences of the videos. People who already felt popular and liked in their own life tended to look at the people nodding, chatting, and smiling; while the people experiencing isolation and loneliness barely noticed any signs of warmth. Instead, they were much more likely to focus on expressions of unkindness or rejection.[24] As the psychologist Mitch Prinstein notes: "It was as if they had watched a completely different movie altogether—focusing far more intently on cues that were barely noticed by others at all."[25]

You may have experienced this yourself before a particularly difficult event, like public speaking: because of our fears, the audience can seem to be full of bored or judgmental faces. Or perhaps you simply wake up in a bad mood and notice that everyone on the train to work appears especially unfriendly that morning. These are temporary distortions. For many people, however, the expectation of hostility can become deeply embedded from a young age—with past rejections casting a long shadow over their entire social world, so that they never truly experience the expressions of friendliness around them.

In each of these examples, the warped view of the world appears totally objective. Thanks to the interaction between our mood, the brain's predictions, and the real sensory inputs, an anxious or depressed person really is "seeing" the world as a much more threatening place, in just the same way that the witnesses at Gatwick "saw" the drones. And this biased processing can have real behavioral consequences, leading you to avoid the very situations that could help to realign the brain's predictions. If an escalator seems much higher than it really is, you are going to find it much harder to place your foot on the first step; and if every face around you appears to be frowning, you will be much less likely to strike up a conversation with someone sitting next to you.

Fortunately, you can learn to neutralize these micro-illusions with training.[26] Indeed, exposure therapy—in which people are encouraged to directly confront their fears—may work by recalibrating people's perceptions. In 2016 a team of German researchers asked arachnophobes to don a virtual reality helmet and wander through rooms containing lifelike representations of spiders, with a basic goal: to stay calm and to avoid running away from the threat. Not only did the participants' fears of real spiders fall over the course of the session, but their estimates of the spiders' sizes also became far more realistic.[27]

You can also target the distorted perceptions directly, using a technique known as "cognitive bias modification." People with anxiety, for example, are given straightforward computer games in which they are presented with a series of facial expressions—illustrated, for example, as fairy sprites hiding in a mountain landscape. The participant's task is to quickly find the smiling, happy face while ignoring the more hostile expression. (If you are interested in trying this yourself, consider downloading the Personal Zen app, which was developed by researchers at City University of New York and—at the time of writing—offers a free trial on most smartphones.) The aim is to readjust the brain's visual processing so that it no longer accentuates the threatening information in a scene. And many patients report significant benefits from these exercises. Even a single session of a program like Personal Zen seems to bring about short-term changes in people's feelings and behavior—improving, for instance, their performance at public speaking—while more regular training leads to longer-lasting benefits.[28]

The simple recognition of the brain's inherent subjectivity has helped me to cope with dips in my mood. When I feel especially anxious or depressed—and the world around me seems to confirm my fears—I try to account for the fact that my emotions, and the expectations that accompany them, might have biased my perception. Given that negative expectations can also bias our attention, I also make more of an effort to look for actual unambiguous acts of friendliness—essentially replicating the bias modification games in a real-life city.

Needless to say, this strategy is no panacea for serious mental illness, but I find that it often stops me from falling into the spiral of negative thinking that would once have exacerbated and prolonged my low mood. It is just one example of how once we understand the power of expectation we can recalibrate our predictions to experience a healthier and happier view of the world.

## TASTE IS IN THE MOUTH OF THE BEHOLDER

The power of expectation is particularly well known in the world of gastronomy, where marketers and chefs have long harnessed the prediction machine to increase people's enjoyment of their dishes.[29]

In one of the earliest experiments on the top-down effects of expectation on taste, in the 1960s two American scientists looked into people's perceptions of astronauts' meals, such as a chocolate-flavored health shake filled with protein, carbs, and vitamins. Without knowing the origins of the drink, people tended to find the taste rather unappetizing—a poor comparison to typical chocolate milk. When the drink was explicitly labeled "space food," however, the public's appreciation dramatically increased. The exotic name—associated as it was with cutting-edge science—raised expectations and, as a result, acted as a powerful flavor enhancer.[30] We now know that this would have been the direct result of their top-down processing, changing the flavor according to their expectations.

More recently, researchers at MIT approached drinkers in two of the university's iconic pubs, the Muddy Charles and the Thirsty Ear, to take part in a taste test. The participants who agreed were given a sample of a regular beer (either Budweiser or Samuel Adams) and the unfamiliar "MIT Brew." Like the "space food," MIT Brew sounded cutting-edge and exciting—as if it were prepared with advanced technology. Unbeknownst to the drinkers, however, it was identical to the regular brands, except the scientists had added a few drops of balsamic vinegar to each glass.

The idea of beer laced with vinegar may not initially sound appetizing, but the drinkers loved the concoction, with around 60 percent reporting a strong preference for MIT Brew over the other drink. Knowledge of the vinegar did not change that preference, provided it was given *after* the tasting. This was not the case, however, if they were told about the nature of the "secret ingredient" before they had tasted it; then only around 30 percent appreciated the beer's unique mix of flavors compared to the other sample. The effect of their expectation on the experience of the beer's flavor was enough to halve MIT Brew's popularity.[31]

You may have experienced something very similar yourself when you have tasted an expensive bottle of wine; thanks to altered expectations of quality, knowledge of a higher price tag can result in a marked improvement in flavor—irrespective of the actual drink.[32] Changes in appearance can have similar effects. When scientists colored white wine red, participants noted much richer notes in its taste—the traces of "prune," "chocolate," or "tobacco" that are typically associated with real red wines. And the power of expectations is so strong that even wine experts fall for this gustatory illusion.[33]

The effects of our preconceptions are evident in scans of the brain's responses to foods. When participants were given the basic umami flavoring MSG, for instance, along with a single sentence detailing its "rich and delicious taste," their brains showed greater activity in the regions processing gustatory pleasure than the brains of those told that they were receiving "monosodium glutamate" or "boiled vegetable water."[34]

Sometimes the exact same substance can evoke intense pleasure or outright disgust, depending on someone's expectations. A mixture of isovaleric and butyric acid, for example, creates a slightly acrid odor that can be found in two familiar substances: Parmesan cheese and vomit. But your brain will process the same aroma very differently depending on how it is labeled, leading you to either salivate or retch.[35]

These perceptual expectation effects are not really so different from the unscrambling of the images on pages 14 and 15—in each case, the labels are helping to make sense of ambiguous signals that could be interpreted in multiple ways. Given these findings, it's little wonder that people's tastes in food vary so much—depending on their expectations and associations, they may be experiencing completely different things.

If you are trying a new food for the first time, you may apply these findings yourself by reading up on the meal beforehand; by knowing why others enjoy the dish, you'll be priming your brain to make sense of the gustatory signals so that you can more fully appreciate the unfamiliar combination of flavors. This will be especially important if you're traveling and the food is far outside your usual comfort zone. The famously pungent durian, for example, will be far less off-putting if you have been guided to recognize the "overtones of hazelnut, apricot, caramelized banana and egg custard" described by some connoisseurs, rather than the usual comparisons to rotting flesh.[36]

You can use the same principles when you host a dinner party. You may not be able to turn water into wine through thought or prayer, but the way you describe your food will strongly influence the ways you and your guests appreciate it. So be sure to season your dishes with some delicious words as you serve—that verbal garnish may be as important as the actual, physical ingredients. (We'll discover the implications of our expectations for digestion and metabolism—and the prospect of weight loss—in Chapter 6.)

## SUPERCHARGED SENSES

By harnessing the prediction machine, we may even be able to sharpen the overall acuity of our eyes and ears, allowing us to see and hear in high definition. If that seems far-fetched, just consider how the branding of sunglasses or headphones can affect people's visual and auditory abilities. In the early 2010s, a team of Israeli and American researchers asked participants to wear a pair of shades and to then read eighty-four words under the glare of a bright light. Everyone had a pair of the same quality, but the people who had been told they were wearing Ray-Ban sunglasses made around half as many mistakes as those who were told their shades came from a midmarket brand. The participants who believed they were wearing the designer shades also completed the task more quickly, in around 60 percent of the time.

Strikingly, the researchers found exactly the same results in an equivalent audio task using noise-canceling ear muffs that were meant to "filter onerous audio frequencies . . . while assisting in hearing conversations." People who believed they were wearing a more prestigious brand (3M) were better able to hear a list of words above construction noise, compared with participants who thought they'd been given a lower-quality product, when everyone in fact had the same gear.[37]

In both experiments, the participants' trust in the (allegedly) high-quality products led them to believe that they would benefit from heightened perception of the relevant sights and sounds—and that was what they experienced, even though there was no actual difference in the technology. The expectation that they could see or hear better than if they used another brand had apparently altered the participants' visual and auditory processing, leading their brains to work harder to build richer and more accurate simulations from the information hitting their eyes and ears.

The discovery echoes a study by Ellen Langer at Harvard University, which found that people's beliefs can have a striking effect on their long-distance vision. The participants were cadets from MIT's Reserve Officers' Training Corps. They first took a standard eye test, providing a baseline for their sight, before entering a flight simulator. Despite it being a computer simulation, they were asked to treat the exercise as seriously as possible: to imagine themselves in a real cockpit and to react in the same ways that a real pilot would. During the subsequent game, four planes approached from the front, and the cadets were asked to read the serial numbers written on their wings. Unknown to the cadets, this was another hidden sight test—the sizes of the numbers on the plane wings were equivalent to the four lowest lines on a standard eye chart.

Langer suspected that the cadets would associate the experience of piloting a plane with having exceptional eyesight, and that this, in turn, would improve the acuity of their vision during the simulation—and that was exactly what she found. Overall, 40 percent of the group could correctly read smaller text (on the sides of the plane wings) than they had been able to perceive in the standard eye chart. A control group, who had not been through a full flight simulation but were instead presented with static images of the numbers on the wings, showed no improvement at all.

To confirm the effect, Langer conducted a second experiment, in which she asked participants to perform some jumping jacks—an energetic exercise that, she said, might improve their sight. Although the movements were unlikely to have changed the optics of the eye over such a short period, these participants once again performed better on a subsequent test of visual acuity, thanks to the belief that athletes have clearer vision.

For a final confirmation, Langer reversed the order of the eye chart, with the smaller letters at the top and the bigger letters at the bottom. She found that the participants were able to read smaller letters than they had been able to on the standard chart, apparently because they had built up the belief—throughout years of previous examinations—that it was easier to read lines that were placed higher.

In each of Langer's experiments, the expectation of better vision boosted the brain's visual processing, leading it to sharpen the slightly blurred images of the letters on the retina.[38] Strikingly, many of these people already had good vision—they were boosting their sight beyond 20/20—but even those with poorer eyesight showed significant improvements.

Don't throw out your spectacles or contact lenses just yet: such mental shifts almost certainly can't make up for a serious optical deficiency. (Short-sightedness is typically caused by a misshapen eyeball, and there is no evidence that this apparently permanent anatomical change is the product of our mind.) But Langer's results suggest that the adoption of certain expectations could at least improve your vision with the lenses you currently wear, ensuring that you see the world as sharply as possible.

Throughout this book, we will find that we are often poor judges of our own abilities, and that it is possible to push the limits of what we can achieve through a simple change of mindset.

## MULTIPLE REALITIES

In her autobiographical novel *Seduction of the Minotaur*, Anaïs Nin beautifully describes the mismatched perceptions of the protagonist, Lillian, and the painter, Jay.

"Lillian was bewildered by the enormous discrepancy which existed between Jay's models and what he painted," we are told. "Together they would walk along the same Seine river, she would see it silky grey, sinuous and glittering, he would draw it opaque with fermented mud, and a shoal of wine bottle corks and weeds caught in the stagnant edges." Jay, Nin writes, was a "realist," intent on portraying the world as objectively as possible. But was his perception actually any more realistic than Lillian's? "We do not see things as they are, we see them as we are," Lillian concludes, in one of Nin's best-known lines.

Our new understanding of the prediction machine reveals the profound truth in this statement, across the breadth of human experience. At the most extreme, expectations can shut down vision completely—as we saw with a patient like Sara. At other times, they will create the perception of something that is not there. And on a day-to-day basis, our preconceptions will alter what is already in front of us—transforming the taste of a food, the emotion written on a face, or the sight of the Seine. These subtle expectation effects may be less dramatic than extreme hallucinations, but as we have seen, their consequences can be substantial, creating either vicious or virtuous cycles in our daily lives. To build on Nin's observations: what we feel and think will determine what we experience, which will in turn influence what we feel and what we think, in a never-ending cycle.

This knowledge will be essential as we turn inward to explore the influence of expectations on our physical health in the following chapters. The prediction machine receives many inputs from inside the body, including "nociceptor" nerves that respond to damage—or the possibility of damage—to our organs and contribute to the sensation of pain. Our expectations will influence the processing of those signals—it can tune them up or down—in just the same way that expectations change our experiences of sight, sound, smell, taste, and physical touch. Sometimes faulty predictions might even create the illusion of pain out of nothing at all; or someone's shifted expectations might cause the agony of a real physical wound to mysteriously vanish.

Even more mysteriously, however, the brain's simulations can also produce measurable physiological changes. As we shall see, our subjective expectations can become our body's objective reality—thanks to the awesome power of the prediction machine.

## HOW TO THINK ABOUT . . . THE SENSUAL WORLD

- Question your own objectivity as an eyewitness. The brain's simulations of the world around you are often right but sometimes wrong—and the humble knowledge of this fact could help you to recognize illusions when they occur.
- If you have a phobia, remember that your brain may exaggerate the threat—so it seems physically bigger and scarier than it really is. Exposure therapy may help you to shrink this perceptual bias.
- If you have anxiety, consider downloading an app—such as Personal Zen—that aims to reconfigure your attention with regard to threats in your environment.

- Whenever you are having a bad day, try to consider the ways your mood and the resulting expectations could bias your view of events. Some situations are unquestionably bad, while other events are more susceptible to expectation effects. Learning to separate the two could prevent you from descending into overly negative thinking.
- Boost your enjoyment of sensory experiences such as meals with the power of language. The way we label foods affects their taste, so think of or look for sumptuous descriptions of the dishes that you are serving to yourself and your guests.

(Recognize this bulldog? It's the original of the high-contrast image on page 15.)

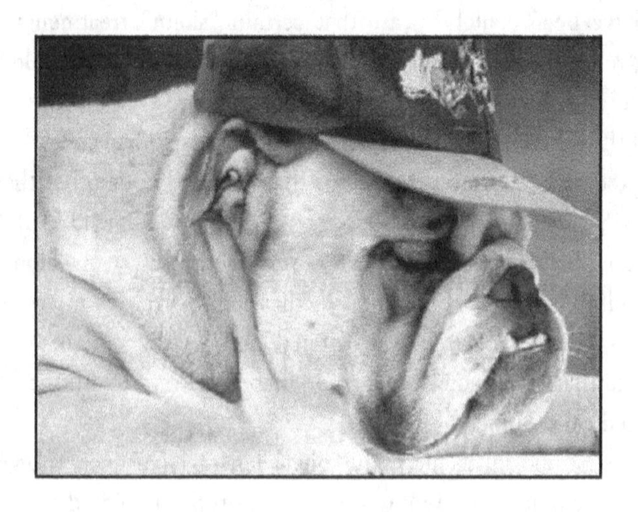

# 2

# A PIOUS FRAUD

## How beliefs can transform recovery from illness

Few scientific ideas have generated as much excitement—or as much outrage—as the placebo effect and the potential of the mind-body connection.

Since the birth of modern medicine in the eighteenth century, doctors have been acutely aware that certain "sham" treatments could bring relief through the patient's beliefs about them. But could these fake cures ever heal the underlying problem? And even if they worked, didn't the inherent dishonesty break the doctor's ethical code?

These were enormous questions that troubled none other than the third US president, Thomas Jefferson. Writing to a friend in 1807, he expressed his fear that some doctors were becoming overzealous with their administration of common medications—such as mercury and opium—which, he feared, were often doing more harm than good. He believed that many complaints could be better served with the *illusion* of a medical treatment.

"One of the most successful physicians I have ever known has assured me that he used more bread pills, drops of colored water, and powders of hickory ashes, than of all other medicines put together," Jefferson said. The deception may have seemed morally dubious, but it was preferable to overprescribing potentially toxic substances that brought about no further improvement for the patient. It was, Jefferson said, "a pious fraud."[1]

Over the following decades, however, doctors became much more skeptical about the benefits of belief. Placebos may bring emotional comfort, they thought, but dummy pills and potions were of little interest to modern medicine—grounded as it was in biological understanding. For some practitioners, placebos were better seen as a diagnostic tool to identify potential malingerers and hypochondriacs: if you found relief from a sham treatment, you didn't have a real disease. By the mid-twentieth century, articles in medical journals derided placebo responders as "unintelligent," "neurotic," "ignorant," and "inadequate"; the *Lancet* described the placebo effect as a "humble humbug." Why would anyone spend time researching such an inane phenomenon?[2]

As a result of this continued skepticism, scientific understanding of the placebo effect has taken a long time to blossom, but we now know that positive expectations can bring much more than emotional comfort, producing real relief for many physical conditions, including asthma, Parkinson's, heart disease, and chronic pain. Even more astoundingly, the cure often produces the same biological changes as the actual drugs prescribed to treat these ailments. The mind-body connection is real and potentially powerful.

The fact that we have evolved this remarkable gift for self-healing is mysterious enough, and its evolutionary origins are the source of much debate among scientists. And there are other puzzles—including the fact that the placebo effect becomes more powerful over time. How can a sham treatment, which is by its very definition "inert" and chemically "inactive," work at all, let alone increase in potency? There is in fact increasing evidence that people can respond to a placebo even when they know they are taking a sham treatment—a finding that seems to defy reason.

The solution to these enigmas, we shall see, comes from the flourishing understanding of the brain as a prediction machine that we explored in the previous chapter. And this new research is inspiring some truly groundbreaking strategies to harness all the benefits of positive beliefs without any deception at all. Dummy pills, it seems, are just one way of packaging an expectation effect, and you can rethink your illnesses and accelerate your eventual recovery using other very simple strategies. With the overprescription of many medications an increasing worry, the use of these psychological techniques could not be more urgent.

More than two hundred years since Jefferson extolled the power of bread pills and colored water, we can tap into the mind-body connection with no need for any kind of fraud, pious or otherwise.

## BELIEVING IS BEING

The renewed interest in the placebo effect began with an American anesthesiologist called Henry Beecher. Serving in Italy and France at the end of World War II, he often had to treat soldiers with truly horrific wounds—torn flesh, shattered bones, and shrapnel embedded in the head, chest, and abdomen. Yet he was puzzled to observe that many of his patients—around 32 percent—reported feeling no pain at all, while a further 44 percent experienced only slight or moderate discomfort. When offered the chance, three-quarters of these people even refused the option of painkilling drugs. For Beecher, it seemed that the relief of having been saved from the battlefield had created a kind of euphoria that was by itself enough to numb their injuries. The patient's interpretation of his illness had somehow allowed the brain and the body to release its own natural pain relief—a phenomenon that could not be explained by the medical knowledge of the time.

According to some reports, Beecher's realization proved to be a godsend, since morphine was in short supply, and soldiers sometimes had to undergo surgery without painkillers—whether they liked it or not. To create the illusion of treatment, Beecher's nurse would reportedly inject the patient with a saline solution while reassuring him that he was receiving the real painkiller. The soldiers often responded surprisingly well to this treatment. Indeed, Beecher estimated that the placebo was about 90 percent as effective as the actual drug; it even seemed to reduce the risk of cardiovascular shock that can result from surgery without sedation and analgesia, and which can be fatal.[3]

These soldiers, who had risked life and limb for their country, were hardly the stereotypical malingerers or neurotics who were normally thought to respond to placebos; nor could you argue that their war wounds were some kind of imaginary disease. Evidently the placebo response was more prevalent and more interesting than was commonly believed. Beecher marveled at the power of expectation to improve symptoms, but he was more concerned about the implications for the testing of new treatments. Active drugs are expensive, and many have undesirable side effects—so you would want to be sure that they are at the very least more effective than a sugar pill or an injection of saline solution.

Beecher's research would ultimately lead to the widespread use of the placebo-controlled clinical trial, in which patients are randomly assigned to take either a fake drug (such as a sugar pill) or the actual treatment under consideration. Neither the doctors nor the patients know which one is being administered until the end of the trial, when the study is "unblinded." Once all the data has been collected, the scientists compare the effects caused by the placebo and the effects experienced by the patients receiving the actual treatment. Only those treatments that significantly outstrip the placebo gain approval.

By the 1970s the US Food and Drug Administration was on board with this protocol, and the placebo-controlled clinical trial was soon recognized as the gold standard of medical regulation. It has been an undoubted benefit to patients: it ensures that they are given provably effective treatments, and it also allows scientists to check the safety of drugs before they are given to the wider population.

Unfortunately, this setup still frames the placebo response as a nuisance; as long as the drug performs better than the sham treatment, the placebo response is often ignored rather than exploited. But the effects are at least recorded, providing ample data for researchers interested in the role of expectation in medicine. And their discoveries over the last couple of decades have been truly remarkable.

Take the potent painkilling effect that Beecher had observed on the battlefield—a finding that has been replicated again and again in placebo-controlled clinical trials of analgesics. Overall, researchers estimate that the placebo response can account for 50 percent of the pain relief afforded by the actual drug.

As we saw in the last chapter, this pain relief could be the result of shifts in subjective experience, as the prediction machine recalibrates its expectations of suffering. Yet it also appears to coincide with distinct physiological changes that mimic the action of the drugs themselves. When people take a placebo to replace morphine, for instance, the brain starts producing its own opioids that can soothe pain. To prove this, scientists administered a placebo painkiller along with the chemical naloxone, which is used to treat morphine overdose by blocking the brain's opioid receptors. Sure enough, the naloxone drastically reduced the placebo pain relief in much the same way that it would have reversed the effects of the real drug. That reaction would be impossible if the pain relief was purely subjective.[4] Instead, it seems that the brain has its own "inner pharmacy," allowing it to create its own medications, like opioid painkillers, on demand.

Equally breathtaking benefits can be seen in treatments for Parkinson's disease. The illness is caused by a deficit of dopamine in the brain. Besides being involved in feelings of pleasure and reward, dopamine is essential for the smooth coordination of movements, which is why patients with Parkinson's often suffer from uncontrollable shaking. The drugs that treat the disease increase levels of dopamine, or act as a substitute for the neurotransmitter by stimulating the parts of the brain that normally respond to it. That reaction would seem to be impossible to achieve with an impotent sugar pill. Yet various trials have shown that a placebo treatment can improve the symptoms of Parkinson's patients by around 20 to 30 percent.[5] Once again, the expectation of improvement somehow allows the brain to mine its own "inner pharmacy," increasing the brain's natural dopamine supply.[6]

Besides changing our brain chemistry, placebos can tune the immune system. Allergies, for example, are caused by an overzealous reaction to typically harmless substances that the body mistakes for a dangerous pathogen. Certain drugs can calm that response—and so can the mere expectation of relief. This can be seen in itching and the size of the welts on subjects' skin, which are significantly smaller in people receiving a sham treatment that ostensibly suppresses the inflammation, compared to those receiving no treatment at all.[7] For people with asthma, meanwhile, an empty inhaler provides about 30 percent of the benefits of the drug salmeterol.[8]

A placebo effect may even explain the benefits of certain forms of surgery, such as the fitting of arterial stents. The procedure involves sliding a catheter through an artery to the area of blockage. When that is in place, a little balloon covered in a wire mesh is guided down the catheter. The balloon is then inflated to widen the artery, leaving the wire mesh—the stent—in place to hold the artery walls open.

This surgery is often essential in medical emergencies such as a heart attack (a situation in which a placebo is unlikely to be of immediate assistance). But stents are also used to ease blood circulation in patients suffering angina, with the aim of reducing ongoing pain and discomfort—and here the role of expectation may be far more significant.

This possibility has come to light only very recently, since, unlike in drug development, doctors and scientists studying surgical interventions are not always obliged to conduct placebo-controlled trials for new operations. Instead, they can use other comparisons—such as "treatment as usual"—which may not produce the same expectations as the new procedure. To find out if a placebo effect could explain some of the benefits of arterial stents, a team of cardiologists from hospitals across the UK split 230 patients into two groups, with half receiving the full surgery and the other half receiving a "sham" surgery, in which the catheter was guided in and out of the artery *without* the stent being put in place. (As in the placebo-controlled drug trials, the patients were informed they might not receive an actual stent—and the team made every effort to minimize the lingering effects of the surgery.)

In a paper for the *Lancet*, the research team reported that both groups were capable of greater physical activity after the operation—measured by their performance on a treadmill—and the benefits of the stent over the sham surgery were too small to be considered statistically significant.[9] Needless to say, the finding has been the source of much debate among cardiologists, and ongoing research will need to replicate the finding before the medical guidelines change. But given this carefully controlled study, it certainly seems likely that much of the stent's benefits for angina arise from the patients' expectations of improvement, rather than from the physical change to the heart's plumbing.

In some cases a placebo treatment could even be a lifesaver. In one trial of beta-blockers, participants who regularly took a placebo were half as likely to die during the study as those who were less diligent in taking pills. Clearly, the placebo wasn't as effective as the active drug, if both are taken at an equally high rate—but so-called "placebo adherers" nevertheless lived longer than people who took pills—either the active drug or the dummy ones—only haphazardly.[10]

The longer life expectancy of placebo adherers has now been demonstrated in many other studies, making it extremely hard to disregard as some kind of statistical fluke.[11] One explanation is that the high compliance simply reflects a healthier lifestyle in general. But the differences remain even when you control for all kinds of variables—such as income, education, and whether someone smokes, drinks, or overeats—that would also predict someone's likelihood of dying. This leaves us with the distinct possibility that the regular ritual of taking a pill has itself helped someone to maintain a healthier body, thanks to the hopes of better health that come from taking a potential medication.[12]

Exactly how and why we react to placebos in all these ways is a matter of fierce debate, but many researchers argue that this type of expectation effect comes from at least two sources. The first is a general healing response, an evolved reaction that allows the body to adapt to the presence of immediate threats. When we have just been injured, for instance, we need to feel pain to avoid further damage to the body—making us more cautious in our movement. If, however, we are in safety and being treated for our injuries, pain serves less of a function, so we can afford for it to be soothed. Similarly, inflammation is essential to deal with immediate contact with a pathogen, but it can prevent other processes from kicking in to heal the damage. It is therefore beneficial for the immune system to rein in inflammation when it perceives that you are already on the road to recovery. Anything that reduces fear and anxiety about your illness—including the sense that you are receiving medical care—may produce this generalized healing response, which can be powerful in its own right. Beecher's soldiers seemed to experience something like this—the mere fact of being away from the battlefield relieved much of their pain—but it will also be true whenever any of us receives medical treatment. According to this theory, placebos are a powerful symbol of care that can trigger this response.

Importantly, the prediction machine has also evolved to fine-tune its actions based on particular experiences, through a learning process called conditioning. If you are taking a placebo painkiller expecting it to be morphine, the release of opioids will be far stronger if you have already taken morphine in the past, for example. Similarly, the placebo-triggered release of dopamine will be far stronger if someone has already taken a Parkinson's drug—and a placebo to reduce transplant rejection will be more effective if someone has already taken a relevant immunosuppressant. In each case, the brain is activating the systems that would make the most efficient use of the body's resources based on its previous memories and associations.[13]

With the right kind of messaging, appealing to the right kind of experiences, it may be possible to turn anything into a placebo. Scientists at Columbia and Stanford Universities even managed to persuade students that a plain bottle of spring water was an energy drink containing 200 milligrams of caffeine, and their blood pressure responded commensurately.[14] You may not even need to be physically present to experience the benefits: a team from Switzerland has shown that a placebo administered in a virtual reality environment can reduce pain suffered by a real-life limb.[15]

In general, however, the prediction machine relies on many different cues to determine its expectations, drawing on a wealth of associations learned in other areas of life, and this means that certain forms of placebo are consistently more potent than others.[16] These factors can be as superficial as size. Many people assume that bigger means better, so if people are taking a pill, a larger tablet may render a bigger response than a smaller tablet. The form also makes a difference—capsules seem to be more effective than tablets—and so does the price. Labeling a Parkinson's treatment "cheap," for example, *halved* the placebo's benefits, compared to an identical injection that was labeled "expensive."[17]

For similar reasons, a drug's marketing can matter to an inordinate degree; a placebo labeled with a well-known brand of painkiller, such as Advil, may be much more effective than a pill that is labeled generic "ibuprofen." Indeed, in one study, a branded placebo was so powerful that it matched the effects of the active painkiller.[18] That shouldn't be surprising: the participants had seen the brand so often—and heard of its painkilling effects—that they would have had fewer doubts of its effects, whereas a generic drug may feel unfamiliar and of a lower quality.

More generally, injections tend to have stronger effects than medications taken orally, and surgery is better still, perhaps because it is easier to understand and visualize its mechanism, compared to treatments that involve complex chemical reactions. We are also swayed by the age of the treatment; if a drug or medical device has just been approved and is generating a lot of excitement, you may feel a larger placebo response than if the treatment first emerged thirty years ago.[19] Last, but not least, there's the relationship between you and the healthcare provider. The placebo effect will be far more potent if they seem caring and competent.[20]

In an amazingly comprehensive way, the brain, working as a prediction machine, updates its simulations and coordinates the body's responses using any cue that might hone its expectations of recovery. There is now no doubt that expectations can—and do—shape our physical reality.

The million-dollar question, of course, is whether we can harness these expectation effects responsibly. Jefferson may have viewed sham treatments as a pious fraud, but it is against doctors' ethical code to lie to a patient, meaning that the deliberate use of the placebo response in general practice has long seemed out of the question—at least officially. (In practice, the use of placebos may not be so rare: 12 percent of general practitioners in the UK report having given saline injections or sugar pills at least once in their career.[21])

What if that deception were unnecessary? What if we could know that our treatment was a so-called sham, and still get better? It may sound paradoxical, but as we shall now see, knowledge of the placebo effect can itself provoke a healing response—equipping patients with the mental tools to treat themselves.

## THE HONEST PLACEBO

Signs of this deception-free healing response may in fact have been hiding within the medical literature. It's just that no one had thought to look—until pharmaceutical companies started hitting a wall in their search for new treatments.

For decades after the birth of the clinical trial, drug discovery experienced a kind of golden age, with a high proportion of experiments revealing new and effective treatments for various conditions—making Big Pharma more profitable than Big Oil. By the turn of the twenty-first century, however, scientists began to notice that many of their clinical trials were failing at greater and greater rates than before. The failures came so fast and so often that some medical research organizations even feared for their financial future.[22]

With intense data mining, the scientists eventually found an answer. The trials were all perfectly well designed, but the people in the placebo arm of the trial seemed to be getting more and more relief from their pills, which made it much harder to isolate the benefits of an actual drug with a provable, statistically significant difference.[23] If you look at tests of painkillers in the 1990s, for example, the active drugs tended to outperform the placebo by around 27 percent. By 2013, that advantage had dropped to just 9 percent. Crucially, this was caused almost entirely by the increased potency of the sham treatments, which brought about 20 percent more pain relief at the end of the period compared to the beginning, while the active drugs did not see a similar rise. (Apparently, the active drugs had reached an upper limit for possible pain relief.)

If they were running a race, the real drugs had started out way ahead, only for an unlikely straggler to somehow narrow their lead. Compounding this mystery, the strange inflation of placebo power seemed to be concentrated in the United States, while trials in Europe were largely unaffected.[24]

How could this be possible? One potential explanation came from direct-to-consumer advertising in the United States. The constant repetition of TV ads could increase people's expectations about the benefits of any drug under consideration. These heightened expectations could inflate the relief felt by those taking the dummy pills—amplifying the release of the brain's endogenous painkillers, for instance—to such a degree that the effects come to overshadow the additional benefits of the active ingredients taken by the non-placebo group. Countries without direct-to-consumer advertising don't have this constant reinforcement of positive expectations, meaning that the size of the placebo response has remained more stable.

There is, however, an even more intriguing possibility: that the increased potency of the dummy pills has arisen from greater public knowledge of the placebo response itself. This theory comes from Gary Bennett at the University of California San Diego, who was part of a team showcasing the increasing efficacy of placebos in treating pain. He points out that in the mid-twentieth century, most people knew very little about placebos and tended to view them somewhat dimly. If you were in a clinical trial, and you worried that you had been given a dummy pill, your hopes of improvement might not have been high. But the recent interest in placebos and their capacity to produce real physiological effects has changed that notion, with the potential power of expectation recently receiving considerable media coverage. Today, the prospect of receiving a placebo does not seem so unattractive, since many people would expect to receive some real relief whether or not they get the actual drug.[25] And thanks to the mind-body connection, that became a reality—increasing the potency of the dummy pills to such an extent that the actual drugs struggled to compete.

Bennett suspected that media coverage of the placebo effect may be especially common in the English-speaking world—explaining why the increase of placebos' potency is particularly marked in the American trials but not across Europe. To test the idea, he examined huge bodies of digitized texts in English, French, German, Italian, and Spanish. As he had hypothesized, the use of the word "placebo" has risen dramatically in the English-speaking world in recent decades, whereas its usage in other countries has barely increased at all. Importantly, this growing recognition was not limited to academic literature but could also be seen in newspapers, popular magazines, and the scripts of TV broadcasts—outlets where the message is most likely to reach the general public. (Unfortunately, the analysis of the pain medication trials did not provide the necessary data to check if the placebo response is increasing in the UK as well as in the United States—which would provide further evidence for Bennett's hypothesis.)

The idea that the word "placebo" can itself evoke a placebo response may sound absurd. Since the eighteenth century, the whole concept of the placebo effect has centered on the premise that people must believe they are receiving a "real" treatment for there to be any noticeable benefit. Jefferson wrote that the deception was a "pious fraud" because it was completely unavoidable. Beecher himself said that "it does not matter in the least what the placebo is made of or how much is used *so long as it is not detected as a placebo by the subject.*"[26] Yet various groundbreaking studies have shown that a large number of people do respond to a placebo even when they are fully aware that they are receiving an inert pill. According to Bennett's hypothesis, this may be most common in regions where the placebo effect is already well known, but there is now plenty of evidence that "open-label placebos" can be equally potent elsewhere, provided scientists give participants a clear explanation of the brain as a prediction machine with the power to influence the body's responses.[27]

Take, for example, a trial of a novel treatment for chronic back pain, run by the health psychologist Cláudia Carvalho at a public hospital in Lisbon, Portugal—the success of which sent ripples across the global scientific community when it was first published in 2016. The patients were given a bottle that was clearly labeled "placebo pills, take twice a day," containing orange gelatin capsules. Carvalho explained that the pills contained no active ingredient, but that they could nevertheless have powerful effects on the body through processes such as conditioning, and she then showed participants a short video to consolidate the idea. To avoid adding to the participants' existing emotional strain, she also emphasized that they did not need to experience a continually optimistic mood—an unrealistic prospect for someone who is in constant pain—for the placebos to have an effect; it was the simple act of taking the pills regularly that would be essential to the treatment's success.

Three weeks later the impact was stark, with the participants reporting a 30 percent reduction in scores assessing their "usual" and "maximum" pain, a large improvement that was not seen in a control group of patients who had continued as normal without the addition of the open-label placebos. A separate questionnaire also revealed a marked improvement in the participants' day-to-day activities, such as the ability to leave the home or perform physically demanding jobs. Overall, the benefits of the open-label placebos met the typical threshold for "clinical significance"—a 30 percent reduction in symptoms—that you would expect of an active treatment.[28]

Even more amazingly, Carvalho published a follow-up paper in 2020 showing that these benefits had lingered for five years after the original trial ended. The knowledge of the placebo response seemed to have stuck with the participants, increasing their overall capacity to cope with their condition.[29] Carvalho's finding echoes an observation made by many of the scientists I have interviewed for this book—that participants in experiments often find knowledge of expectation effects to be highly empowering, with long-lasting benefits that continue far beyond the initial trial.

Open-label placebos have now proven successful in the treatment of a number of other conditions, including migraine, irritable bowel syndrome, depression, attention deficit hyperactivity disorder, and menopausal hot flashes.[30] They have even helped to soothe the burning eyes, sore throats, runny noses, and itchy skin suffered by people with hay fever.[31] But it is their painkilling effects that continue to generate the greatest excitement, since they offer a potential approach to reducing the opioid addiction crisis.[32]

According to the US Centers for Disease Control and Prevention,

450,000 Americans died of an opioid overdose between 1999 and 2019, many of whom had been hooked on prescription pills.[33] An open-label placebo could reduce people's reliance on those drugs, with the power of expectation helping the brain's natural analgesics to slowly replace the pharmaceutical products as the patients reduce their dosage of the actual opioid. That may seem ambitious, but our knowledge of the placebo effect offers some sophisticated strategies that could maximize the chances of success. You can initially pair an actual drug with a strong and evocative smell, for instance, strengthening the body's response whenever it receives a placebo pill matched to a similar odor.

A recent study led by Leon Morales-Quezada at Harvard Medical School did just that. The participants were all in rehabilitation for severe injuries—such as damage to the spinal cord. For three days they were given a strong opioid along with a clearly labeled placebo pill and asked to sniff a swab laced with the strong smell of cardamom oil at the same time—after which point they were encouraged to forgo the real drug whenever possible.

The results were incredible, surpassing the researchers' wildest expectations. At best, they had expected that the placebo pills might cut down the participants' opioid consumption by around a third. In fact, the patients reduced their intake by 66 percent over the course of the trial—without any increase in their pain or discomfort.[34] The open-label placebo had allowed them to radically cut their dose of these potentially addictive pills without any extra suffering.

The aim now is to build larger, extended trials with the ultimate goal of weaning participants off their pills completely. Anecdotally, Morales-Quezada told me of one patient who managed to go opioid-free within just three days using the same technique. We need more evidence than a single case study, of course. But for the time being, his results offer another exciting glimpse at the potential use of the placebo response to reduce suffering, without the ethical complications that previously troubled doctors. The power of the word "placebo" may be the bane of pharmaceutical companies hoping to invent new drugs, but it could be an enormous boon for many patients looking to avoid the risk of addiction or to overcome its iron grip.

## NO PLACEBO, NO PROBLEM

A few years ago the medical psychologist Johannes Laferton received a postcard from a former patient. It was the kind of message that any scientist would be thrilled to read. "As promised, I send you and your friendly colleague best greetings from Italy. You were so encouraging!" she wrote. "Before surgery, I did not expect to be able to spend my holidays here at this wonderful place. I feel quite well."

Only three months previously the sixty-seven-year-old patient had undergone a heart bypass. Needless to say, the five- to six-hour surgery is often traumatic for the patient, and many people continue to suffer general disability many months after the operation. Laferton was part of Winfried Rief's team at Philipps University of Marburg in Germany, and they hoped to ease the recovery process and maximize the benefits of the operation through invoking the mind-body connection—with no placebo pill in sight.

The trial was known as PSY-HEART, and it involved two in-person meetings and three short telephone calls that were designed to improve the patients' expectations of what was about to occur. During these conversations a psychologist would explain the procedure in detail and describe the ways that it could help to alleviate their coronary heart disease—discussions that might be skipped in a typical consultation but that helped to build up the patients' beliefs in the operation's benefits. The patients were then encouraged to form a personal "recovery action plan," in which they laid out some optimistic—but reasonable—outcomes from the surgery. (For the sixty-seven-year-old patient who sent Laferton the postcard, this progressed from smaller steps, such as gardening, to socializing and traveling with her friends.) They were also taught a visualization exercise and asked to imagine their life six months after the operation.[35]

As a comparison, the Marburg researchers set up a second group who were given more general emotional support in exactly the same number of sessions as those undergoing the new intervention—without explicitly discussing their expectations of the treatment. This control mechanism set a very high threshold, since experiencing empathy and social connection can trigger a healing response. For a second point of comparison, the team therefore looked at the progress of patients who did not receive any additional help at all, but who simply underwent the same procedures offered to typical patients undergoing heart bypass surgery.

Immediate differences among the three groups could be seen in the length of people's hospital stays. On average, the patients with enhanced expectations were discharged around 4.7 days earlier than those receiving the standard medical care, while the people receiving the social support fell somewhere in between.[36] When you consider the cost of postoperative hospital care, that advantage alone should make the intervention an attractive proposition for health services, with savings that easily outweigh the expense of the psychologists' time, which came to about three hours per patient.

The benefits continued to accumulate in the months following the operation. When questioned on the ways that their discomfort affected their family life, recreation, sexual behavior, and sleep, the patients who had been encouraged to develop positive expectations tended to show the quickest recovery. By the end of a six-month follow-up period, they also reported a greater capacity to return to work, compared to the participants who received the emotional support or those offered the standard care.[37]

Importantly, these improvements were not just self-reported but appeared to coincide with objective, biological differences between the groups. For example, the team measured levels of pro-inflammatory molecules, such as interleukin-6 (IL-6). Besides producing general feelings of sickness, these molecules are known to damage blood vessels—potentially reducing the benefits of the surgery and increasing the risk of further heart disease. As Laferton and his colleagues had hoped, the patients with improved expectations tended to show lower levels of IL-6 at the six-month follow-up.

The patients' improvements were probably a combination of behavioral and psychosomatic changes—perhaps through some kind of "virtuous cycle." The enhanced expectations and associated biological response may have made it easier to perform physical activities, which in turn reinforced the positive beliefs about their healing and contributed to further improvements, accelerating their return to a happier and healthier life.

How should we interpret these findings? PSY-HEART clearly builds on research into the placebo effect and appears to work through a very similar mechanism—yet it shows that you can do away with the sham or dummy treatment altogether. Instead, with the help of the psychologists, the participants were reconfiguring the prediction machine's expected outcomes, using rational analysis to resolve unfounded doubts and to set out a realistic vision of the treatment's benefits. That approach may appeal more to patients who shy away from the use of an open-label placebo—which may feel too artificial or fake—but who are nevertheless open-minded about the possibility of rethinking their outlook.[38]

Promisingly, it may be possible to integrate this kind of expectation effect into many other medical procedures. Having helped design the PSY-HEART trial, Keith Petrie and colleagues in New Zealand recently examined whether positive expectations could help people with anemia who were due to receive an iron infusion into the bloodstream. Before they were hooked up to the infusion, the participants were shown a graph demonstrating the changes to expect in their hemoglobin levels, and the reasons that the infusion would increase their body's energy supply. Four weeks later, the scientists gave the participants a standard questionnaire designed to measure their energy levels in daily life, including any potential effects of fatigue on their memory, concentration, and physical activity. Sure enough, they found that patients with enhanced expectations of the treatment showed markedly lower levels of fatigue than a control group, who had not been offered the extra information on the treatment's effects. (These participants had instead discussed practical measures like diet and exercise that could improve health overall.) By changing people's interpretation of the treatment and their expectations of its success, the short conversation about the treatment's mechanisms had maximized the treatment's benefits.[39]

Sometimes a single sentence may make all the difference. It is common, after all, for doctors to see patients with conditions that will naturally ease with time. In these cases, no active treatment is necessary, but the doctor can still accelerate the healing process with the words they say. In one recent study, led by Kari Leibowitz at Stanford University, participants were first provoked with a mild allergic reaction on the skin, leading to an irritating itchiness. The participants then remained in the lab for around twenty minutes afterward. With some of them, the researchers checked the state of the skin reaction without much comment; for others, they explicitly described how the rash and irritation would soon go away. Those reassurances became a self-fulfilling prophecy that soothed the participants' symptoms so that they recovered more quickly than did the control group.[40]

You might hope that this kind of exchange would already be common in medicine. But Leibowitz points out that visits to the doctor are sometimes seen as a waste of time by the patient unless a medicine is prescribed to validate their illness. Her study shows that prescription-less visits *do* have value, since the conversation can reduce patients' discomfort without any medication at all. Leibowitz's findings recall another striking study, which found that patients recovered more quickly from the common cold, with markers of greater immune activity in the nose, if their doctors showed a more reassuring and empathetic attitude in the interactions and stressed the fleeting nature of the infection.[41] (On average they recovered a whole day earlier than the people seeing a less reassuring physician—a significant change, considering that colds rarely last longer than a week.) The words a doctor speaks are themselves "biologically active" and an essential element of any treatment.

Importantly, none of these exciting new treatments involves instilling a sense of false hope. Each project simply used the facts at hand to help the patient understand the process and their prognosis and to frame their progress in the most positive way possible. It's an approach that we will see again and again in the rest of this book. When it comes to the mind-body connection, knowledge really is power.

## THE WILL TO LIVE

Looking back, it's tempting to wonder how medicine might have progressed if researchers had paid more attention to the kinds of effects that Thomas Jefferson noted in 1807—when he first described the use of dummy cures to prevent the overprescription of active drugs such as opium. He may have called it a "pious fraud," but we have now seen how you can prevent opioid misuse without any kind of deception. Honest strategies to enhance patient expectations can and should be an essential element of all evidence-based medicine.

Jefferson did not write on the subject of placebos again. There is, however, another reason why he is of interest to researchers studying the mind-body connection—and that is the specific day he died. The president's health had started to deteriorate in 1825, with a series of intestinal and urinary disorders. By June 1826 he was completely bedridden, beset by a terrible fever, yet he survived until July 4—the fiftieth anniversary of the signing of the Declaration of Independence.

Astonishingly, Jefferson's presidential predecessor John Adams also died on the same day in 1826. Having not yet heard the news of his rival's passing, his last words were apparently "Thomas Jefferson survives."

Is it merely a coincidence that the second and third presidents of the United States died on this landmark occasion? Or had something more interesting occurred? John Adams's son, John Quincy—who was the sitting president at the time—described the timing of the two presidents' deaths as the "visible and palpable remarks of Divine Favor." Scientists don't generally believe in that kind of godly intervention, so they have looked for other answers. And they have argued that the timing of the deaths may have resulted from a psychosomatic effect. Perhaps in their old age the two former presidents experienced a great will to survive until the momentous anniversary of the country they had helped to found—and once they had reached that day, their bodies quickly gave out.

It may sound fantastical, but as we shall discover in the next chapter, the prediction machine has a dark side that, among many other important consequences, means that our thoughts and feelings really can determine the time of our death.

## HOW TO THINK ABOUT . . . HEALING

- If you hear that a treatment's effects can be partially explained by the placebo effect, don't panic! Remember that the biological effects are still meaningful, even if they arise from expectation.
- If you have a choice over medical treatment options, try to bear in mind the factors that can influence the size of the placebo component. All other things being equal, bigger tablets

are more effective than smaller tablets—but capsules are better still.

- Similarly, if you have a choice over your health-care provider, try to choose someone you find to be empathetic and caring. Their attitude could change the way you respond to the treatment they offer.
- Try to find out—from your health-care provider, or from another credible source—how your treatment works and the way it might bring about its benefits. That knowledge can strengthen the effects of the treatment.
- Based on this information, try to visualize your recovery and, if relevant, make a plan for your journey to better health. Doing so should maximize your chances of improvement.
- If possible, meet other patients who have benefited from the same therapy and are willing to share their experiences. These conversations could help to shift your expectations of the treatment and its success.
- Consider obtaining an open-label placebo (they are commercially available from certain online retailers). You should never use this as a replacement for actual medications without the advice of a medical professional, but taken along with your existing treatments they may enhance the benefits.
- Above all, be realistic but optimistic about what the mind-body connection can achieve.

# 3

# DO NO HARM

## How expectations can hurt as well as heal—and how to break a curse

Browse the American Psychological Association's dictionary and you'll find a mysterious entry for "bone-pointing" syndrome. The term refers to a tradition of the Aboriginal societies near the red sand hills of central Australia. According to anthropologists visiting the groups in the mid-twentieth century, a shaman could dish out a deadly punishment by pointing a human or kangaroo bone at the wrongdoer and chanting a curse. Almost immediately the victim of the curse would become despondent—and as the curse took effect, the body would weaken, and within days it would fail completely. It was, according to one shaman, a "spear of thought" that killed a person from within.[1]

Similar reports of "voodoo death" can be found across the world.[2] And as we also saw with the Sudden Unexpected Nocturnal Death Syndrome of the Hmong immigrants in the United States, people reporting on these phenomena have often assumed that people in "scientific" societies would be immune to deadly expectations. (The American Psychological Association explicitly describes voodoo death as a "culture-bound syndrome," unique to specific populations, rather than a condition that is universal to humanity.)

The historical and medical literature, however, tells a different story.[3] Consider the notorious case of a man from Nashville, Tennessee, who was diagnosed with esophageal cancer in the 1970s. Surgeons successfully removed the tumor, but further scans revealed that the cancer had spread to his liver. He was told he would be lucky to live until Christmas that year. In the end, he survived to celebrate the holiday with his family but only just: he died in early January.

The man's fate would have seemed like yet another tragic loss to a terrible illness, except an autopsy revealed that the original diagnosis had been wrong: he had a tumor on his liver, but it was small and operable—it could not have killed him. Could his own doom-laden thoughts have led to his death? That was the conclusion of his doctor, Clifton Meador, who described the misdiagnosis as a kind of "hex."[4] For this poor man, the dread of cancer appears to have provoked a response that was remarkably similar to a paranormal curse.

The Australian oncologist G. W. Milton came to similar conclusions when diagnosing people with skin cancer. "There is a small group of patients in whom the realization of impending death is a blow so terrible that they are quite unable to adjust to it, and they die rapidly before the malignancy seems to have developed enough to cause death," he wrote.[5] Knowing about Aboriginal traditions, he claimed that "self-willed deaths" were simply another example of the "bone-pointing syndrome" of Aboriginal societies.

Many scientists now believe these anecdotes represent an extreme version of an expectation effect known as the nocebo response.[6] As we learned in the introduction, placebo means "I shall please," and nocebo means "I shall harm"—and the nocebo response occurs when we believe the body to be under threat. Through the actions of the prediction machine, such expectations will change our physiology so that the mere thought of having a symptom or disease can make us ill.

Death by expectation may be the most extreme example, but the nocebo effect is responsible for many other forms of suffering in our daily lives. It can exacerbate the symptoms of allergies, migraines, backache, and concussion; indeed, whenever we are unwell, a nocebo effect will make our sickness worse. Negative expectations can also contribute to the nasty side effects of the drugs that are meant to cure us of our ailments—and they are a major reason for people to discontinue their medications.

Thankfully, our new understanding of the brain as a prediction machine provides us with innovative strategies to mitigate those effects and neutralize our self-made curses. Combined with the reframing techniques that we explored in the last chapter, these methods should provide some much-needed relief for all kinds of pain and discomfort.

## TOXIC THOUGHTS

Like the placebo response, the potential power of negative expectation has been known since the earliest days of modern medicine, long before the nocebo response even had a name.

The surgeon John Noland Mackenzie was among the earliest pioneers to examine a negative expectation effect in medicine. While working at Baltimore's Eye, Ear, and Throat Charity Hospital in the 1880s, he was asked to examine a thirty-two-year-old woman with severe asthma combined with terrible hay fever. When she was exposed to pollen, her nose and eyes would run and her throat would itch so badly that she felt she must "tear [it] out with her nails"; on the worst occasions she would suffer hourlong sneezing fits. The attacks were so uncomfortable that she had to spend much of the summer in bed, and it was impossible for her to keep flowers in the house. Even the sight of a distant hayfield was enough to set off her paroxysms.

Mackenzie doesn't say what sparked his skepticism, but something about the woman's descriptions led him to question the role of pollen in her symptoms. To test his hypothesis, he acquired an artificial rose "of such exquisite workmanship that it presented a perfect counterfeit of the original." Before the patient arrived, he carefully wiped every leaf and petal, removing any grains of pollen that could possibly trigger an attack.

The patient turned up in surprisingly good health, and after an initial examination and breezy chat, Mackenzie casually revealed the artificial rose from behind a screen. Her resulting distress could not have been greater if he had revealed a bunch of real flowers: her voice became hoarse, her nose blocked, and she had the irresistible urge to sneeze—all within a minute of seeing the object. Examining the woman closely, Mackenzie found visible irritation in her nose and throat, which were red and swollen—she certainly wasn't faking the effects. In this peculiar case, Mackenzie concluded that the "association of ideas" appeared to have been as potent as actual pollen granules.

Needless to say, the patient was amazed to discover the true nature of the rose, and she had to inspect it in detail before she was fully convinced that it was not a real flower. Despite her initial incredulity, the realization brought a happy ending to her illness without any further treatment. On her next visit to the hospital, she buried her nose in a large bunch of real roses without a single sneeze.[7]

A handful of equally ingenious studies revealing the power of negative thoughts followed piecemeal over the subsequent decades. It was only with the rise of clinical trials in the 1960s and 1970s, however, that research into these negative expectation effects began to intersect with placebo research. People's beliefs about a sugar pill, scientists discovered, could both heal their existing symptoms *and* create new noxious side effects that mimic adverse reactions to real drugs—often at the same time.

Henry Beecher, the war doctor and anesthesiologist we met in the last chapter, had in fact noted this possibility in 1955 with his influential paper on the "powerful placebo." Drawing on a few existing experiments, he reported that patients receiving the dummy pills often experienced symptoms such as nausea, headache, dry mouth, drowsiness, and fatigue—all the kinds of reactions that people might report when taking an actual drug. In one trial of an anxiety medication, a member of the placebo group even developed a diffuse rash that cleared up only after they stopped taking the inert pills; another reported palpitations; a third had serious diarrhea within ten minutes of taking the pills.[8]

More than six decades later, we know that this phenomenon is worryingly common. A team of researchers from Oxford, Cardiff, and London recently analyzed data from more than twelve hundred placebo-controlled trials. They found that *around half* the people receiving the dummy pill reported at least one "adverse event" in the average trial. And in 5 percent of cases these reactions were so severe that the participants discontinued treatment altogether.[9] Some of these symptoms may have been misattributed and due to completely unrelated sources of discomfort, but a significant number of symptoms appear to have arisen from the doctors' and drug companies' warnings of certain side effects—suggesting a highly specific expectation effect.

Take an investigation, published in 2007, of finasteride—a medication that is often used to treat men with an enlarged prostate. The drug has been known to result in erectile dysfunction and reduced libido, side effects that feature prominently in leaflets and health websites.[10] To find out if that information could be exacerbating men's frustration, a team from the University of Florence set up a yearlong trial in which half the participants were specifically warned about these possible side effects, while the other half were not. They found that the explicit warnings increased the prevalence of erectile dysfunction from around 10 to up to 30 percent—a threefold increase in a quality-of-life-changing symptom caused by a single piece of information.[11] Exactly the same patterns could be seen in people taking aspirin to calm angina. They were six times more likely to discontinue their treatment due to increased nausea and indigestion if they had been warned of irritation to the stomach and intestines.[12]

The nocebo response seems to be especially important when we are feeling pain. You may have experienced this yourself when undergoing small medical procedures. How often does a doctor or nurse warn that "this may hurt" before giving an injection or taking a blood sample? The thinking behind these words may be that it is best to allow the patient to steel themselves for the pain. In reality the short statement will make that pain more likely. Women receiving an epidural, for example, were told in one study that "you are going to feel a big bee sting; this is the worst part of the procedure." They reported much more discomfort than another group who were reassured that they would remain comfortable throughout the procedure.[13] When someone is warned of pain, you can observe notable differences in the signaling of the spinal cord and brainstem—changes that would be highly unlikely if the participant were deliberately exaggerating the effects for sympathy.[14]

Nocebo responses can be so strong that they overpower the would-be positive effects of active drugs. A numbing analgesic cream can make participants feel *more* pain if they have been told to expect greater sensitivity, and that sensation is accompanied by a rise in blood pressure that appears to signal their distress. Similarly, a muscle relaxant can make people feel more tense if they are told it is in fact a stimulant.[15]

Exactly how the prediction machine brings about these effects is a matter of ongoing research, but in many cases it appears to be the direct reversal of the placebo response—a kind of evil mirror image of all the physiological changes we saw in the last chapter. Whereas positive expectations can trigger the natural release of dopamine and opioids, for example, our negative expectations can deactivate these same neurotransmitters.[16] To make matters worse, negative expectations of pain can trigger the release of chemicals that actively increase our discomfort, such as the hormone cholecystokinin (CCK), which boosts the transmission of pain signals[17]—the equivalent of attaching our nerves to a public address system, ensuring that painful messages dominate everything else. Based on its expectations of illness, the prediction machine will also instruct the nervous, immune, circulatory, and digestive systems in certain ways that could result in inflammation, altered blood pressure, nausea, and the release of hormones that will further amplify our stress.

Since the prediction machine draws on its memories to plan its responses, your chances of experiencing a nocebo side effect will depend on your personal history. If you have suffered a bad reaction to one drug, you are much more likely to experience the same side effects with another treatment, even if it works through entirely different mechanisms—and even if it is a dummy pill.[18] This situation is similar to the common experience of associating sickness with certain foods. If you happened to have a stomach bug after eating a certain meal, for years afterward the very thought of that dish could leave you feeling nauseated—thanks to the overprotective prediction machine preparing you for another onslaught.

As we saw with the placebo response, our expectations of events, and the resulting experience of symptoms, can also be swayed by superficial factors. People may experience fewer side effects from branded rather than generic drugs, for instance, perhaps because their slicker marketing increases the patient's trust in the medication.[19]

Even tiny alterations to a drug's appearance can lead to a huge spike in adverse reactions, a fact that GlaxoSmithKline discovered to its great cost in the late 2000s. For decades, tens of thousands of New Zealanders had been using a thyroid hormone replacement drug called Eltroxin, with just fourteen complaints of adverse events across thirty years. In 2007, GSK decided to move the tablet's production to a new factory, which necessitated a change in the tablet's formulation, resulting in a slightly altered appearance—from yellow to white—and taste. The active ingredient was still exactly the same; the pharmaceutical company had simply altered the binding ingredients that bulk up the pills, and extensive testing revealed that the drug was absorbed and metabolized at the same rate. Patients should have been able to carry on with their treatment without even noticing the difference.

Unfortunately, that reassuring information did not reach patients in time, and many assumed that the altered appearance was a sign of cost-cutting and poorer production. As the pharmacies started to stock the new pills, reports of totally new side effects, including headaches, rashes, itchy eyes, blurred vision, and nausea, began to roll in. Needless to say, concerns soon reached the local media, which jumped on the story. Within eighteen months they had 1,400 new reports of side effects—a roughly two thousand–fold increase on the previous rate of one report every couple of years.[20] It took many more months for the fears to die down and the number of reported adverse events to return to their previous level.[21] Lest we think something about New Zealanders made them uniquely susceptible to this nocebo effect, a very similar health scare involving a reformulation of the same drug engulfed France a few years later.[22]

If you do not currently take any medication, you might assume that you would be immune to negative expectation effects, but there are many other ways that your health could be affected by a nocebo-like response. We all have different "illness beliefs" that change how we interpret our bodily sensations, and these thoughts can have important consequences for many common ailments. The neuroscientist Gina Rippon, for instance, argues that the experience of premenstrual syndrome can be influenced by expectation. In one study, participants were given sham feedback about their position in the menstrual cycle, and that false information turned out to be a better predictor of the symptoms the women reported than their actual hormonal status.[23]

Motion sickness shows a similar pattern; for many it is the expectation of discomfort that brings about nausea during travel, rather than the actual movements of a vehicle—and altering people's beliefs about their own susceptibility can miraculously settle their stomach.[24] The same is also true of the lingering symptoms of injuries such as whiplash, back pain, and mild concussion: there is strong evidence that negative expectations can prolong people's suffering.[25]

One study of mild traumatic brain injury, for example, found that people's initial beliefs about their future prognosis successfully predicted the risk of their actually developing post-concussion syndrome in 80 percent of cases. Indeed, the patient beliefs proved to be a better indicator of lingering distress than the severity of the patients' symptoms at the time of the impact.[26] If you think that your symptoms will last for a long time and are beyond your control, they are far more likely to stay that way, all other things being equal. (The role of expectation should not be a reason to take a laissez-faire attitude to these injuries, of course—the fact that nocebo effects can worsen and prolong the symptoms does not make concussion any less of a problem, and the victims are in no less need of medical support.)

Illness beliefs often vary among countries—a fact that can explain some puzzling geographical variation in people's symptoms. A comparison of mild head injuries in North America and Eastern Europe found that Canadians' post-concussive symptoms (such as dizziness or fatigue) linger for months longer than the symptoms of comparable injuries in Greeks and Lithuanians, and this discrepancy appears to reflect the underlying expectations of each population.[27]

There is a danger that nocebo responses will be mistaken for hypochondria, but that is a terrible misreading of the science. In many cases, people's symptoms will have started with a physical trigger, the effects of which are amplified and prolonged by a nocebo response. For others, the cause may be purely psychological, but that doesn't make the symptoms any less serious. As Mackenzie's asthmatic patient showed us more than a century ago, and many careful experiments have confirmed since, the expectations of illness can themselves bring about observable changes to the body that are just as "real" as the effects of a material pathogen. The stark truth is that nocebo responses are an inevitable consequence of the human brain's predictive processing.

Anytime we feel unwell, our thoughts will be shaping our symptoms—and we ignore that fact at our own peril.

## THE HEART OF THE MATTER

How about those "self-willed deaths"? Is death by expectation possible? Over the past few years doctors have documented a couple of extreme nocebo responses that would certainly lend credence to this assertion. And while these dramatic cases may remain a rarity, they reveal fascinating insights into a serious—and currently under-recognized—risk factor for cardiovascular disease that may be affecting many people.

Let's first consider the case of Mr. A, reported by doctors in Minnesota in 2007. Hurt badly by a recent breakup, Mr. A had signed up for the clinical trial of a new antidepressant in the hope that the new treatment could relieve his feelings of hopelessness. Initially he found that the pills worked, helping his mood to improve. The benefits did not last long, however, and in the second month of the trial he decided to end it all, eating all twenty-nine of his remaining capsules. Quickly regretting the decision, he asked his neighbor to drive him to the emergency department of the local hospital in Jackson. "Help me, I took all my pills," he told the staff as he entered—before promptly collapsing.

When the doctors examined Mr. A, he was pale, drowsy, and shaking, with worryingly low blood pressure, and they quickly attached him to a drip. Over the next four hours, his condition failed to improve. Yet there appeared to be no trace of the relevant toxins within his system, so the medical team called in one of the doctors from the clinical trial—who confirmed that Mr. A had never taken the active drug. According to his physiological signs, he had nearly overdosed on dummy pills.[28] Fortunately, on learning the news he made a full physical recovery.

In an equally remarkable case from 2016, a German woman in Greifswald underwent a trial for acupuncture to reduce the pain during and after a cesarean section. To ensure her informed consent, the patient had been told that there was a very small risk the acupuncture could result in a "vasovagal reaction" such as dizziness or fainting—or, in extreme cases, "cardiovascular collapse." Soon after the treatment started the patient began sweating profusely. Her feet and hands went cold, and her blood pressure plummeted to a dangerously low level, with a heart rate of just twenty-three beats a minute. Fearful of these changes, the team immediately put her on a drip and transferred her to the delivery room, where she recovered enough for the cesarean section to be performed—but if the drop in blood pressure had persisted, the patient and her baby could have easily been at risk. Needless to say, it was highly unlikely that the dangerous drop in blood pressure would have arisen from the acupuncture, but the patient had not even received the real therapy. She had been in a control group in which the acupuncturist had merely placed adhesive tape over her body.[29]

The prediction machine appears to somehow have disrupted the body's vital functions to the point of collapse—and for some people that disruption really could lead to death. There are many ways that this could happen. According to one leading theory, such a rapid physiological decline could arise from high concentrations of stress hormones known as catecholamines, which can be toxic for the heart, and which seem to be released under intense emotions. If left unchecked, their spike could lead to an untimely death.[30] That result would, of course, be much more likely for someone with an existing heart condition—but if the response was strong enough, it could even kill those in good health.

Fearful expectations may have a gradual as well as a sudden effect on people's mortality. Consider a now world-famous study in Framingham, Massachusetts, that has tracked the health of thousands of adults since 1948. In the mid-1960s, a subset of female participants were asked whether they were "more likely," "as likely," or "less likely" to develop heart disease compared with other people their age. The researchers found that the women who had answered "more likely" were around 3.7 times more likely to experience fatal cardiac arrest over a twenty-year period than the other members of the study. Importantly, the women had developed and expressed this expectation before any signs of cardiovascular disease had surfaced; given their health at that point, their fears did not seem to have a good factual basis.[31]

Skeptics may wonder whether behavioral differences among the participants might explain this increased risk of death. No doubt lifestyle could have played a role, yet the increased risk held up to scrutiny, even after the researchers had considered many other health factors, including the participants' body mass index, cholesterol levels, smoking habits, and reported levels of loneliness, all of which can damage the heart. For this reason, many researchers believe that the negative expectations themselves had created a physiological nocebo response, that heightened levels of stress hormones and chronic inflammation had harmed the women's health in the long term and directly contributed to their eventual deaths. It is easy to imagine how, if you believe you are at higher risk of heart disease, each day may be filled with doom-laden thoughts, and every feeling of ill health could be interpreted as a sign of your deterioration—thoughts that eventually become a self-fulfilling prophecy.

The possibility of a slow-burning nocebo effect fits with another recent study that looked at people who were already suffering from coronary artery disease. Soon after the onset of their condition, the patients were asked to rate their level of agreement with statements such as "I doubt that I will ever fully recover from my heart problems," or "I can still live a long and healthy life." Regardless of the initial severity of their illness, the patients with the more dismal expectations were significantly more likely to die over the following decade, compared with patients who took a rosier view of their chances of recovery.[32] Once again, it could be that those participants with negative expectations were less likely to take proactive care of their health—a consideration that was not fully controlled for in this study—but the researchers also noted pronounced stress in certain patients, which, they hypothesize, contributed to the higher mortality rate.

We know, after all, that other kinds of highly emotional strain can lead to increased mortality rates. People are about twice as likely to suffer a heart attack or stroke in the thirty days following the death of a spouse, for example, compared to people who have not suffered a recent loss.[33] It's striking that many victims of "self-willed death"—including the Hmong immigrants to America, the Aboriginal bone-pointing victims, and G. W. Milton's cancer patients—appear to have experienced something akin to grief during their decline as they contemplated their own imminent ends.

An understanding of the prediction machine may also explain why people tend to die on personally significant days—which brings us back to the deaths of Thomas Jefferson and John Adams on July 4, 1826. As remarkable as that coincidence may seem, various studies have shown that the risk of mortality throughout the year is not evenly distributed. One analysis of the death records of more than thirty million Americans found that people are more likely to die on or just after a significant occasion than just before a big event. They found that death is 4 percent more likely to happen on a birthday, for example, than on the preceding two days. (Sadly, this phenomenon seems to be more pronounced for children, who presumably attach an even greater importance to the event—and a greater desire to live to see it—than adults do.)

Similar patterns have now been described in many other countries. (In Mexico, a peaceful death on a significant day is even considered to be a *muerte hermosa* or "beautiful death.") And the data seems to rule out some obvious explanations, such as the idea that people may have been more likely to take their own life on these occasions or to be killed in automobile accidents after late-night festivities. Instead, it seems that for a sizable number of people the body may already be ailing, but it is able to hold on until the event, after which expectations of death precipitate a decline, leading to the increase in deaths on those special days. Supporting this hypothesis, one analysis found that the marked rise in deaths around New Year's Day—a surprisingly consistent phenomenon—was especially high on January 1, 2000, compared to other years. It seems natural that the mind would attach huge importance to the (popularly understood as) millennial celebrations, creating a strong desire to live through a once-in-a-thousand-years event.[34]

For Adams and Jefferson, the fiftieth anniversary of American independence would have certainly constituted this kind of milestone. Amazingly, the fifth president, James Monroe, also died on the same date five years later. As the *New York Evening Post* wrote at the time: "Three of the four presidents who have left the scene of their usefulness and glory expired on the anniversary of the national birthday, a day which of all others, had it been permitted them to choose, [they] would probably have selected for the termination of their careers."[35] There is absolutely no reason to think the timing of their deaths was a deliberate choice—but it may have been an unconscious one, reflecting the prediction machine's profound influence over our fates, right up until our last breath.[36]

## BREAKING THE CURSE

So "death by expectation" really could be possible, and fears and anxieties about illness may even contribute to heart disease in a surprising number of people. It's important to remember, though, that the less extreme and more mundane nocebo effects can still have powerful consequences for our everyday health and well-being. The headaches I experienced when taking antidepressant pills, for example, were truly agonizing. That discomfort wasn't going to kill me, but had I not discovered that symptom's potentially psychosomatic origins, it could have easily persuaded me to discontinue a treatment that ultimately proved to be very effective.[37] When you consider the sheer prevalence of nocebo effects and the discomfort they cause, finding a way of neutralizing their power over us would be an amazing advance for medicine. So how can we achieve it?

This question is both a practical problem and an ethical dilemma. Doctors famously pledge to "first, do no harm," but they are also obliged to obtain their patients' informed consent before treatment. These directives can work at cross-purposes. How can doctors honestly explain medical risks without inadvertently inducing a nocebo response? Over the past few years I have been heartened to see that many scientists are already investigating possible solutions to these paradoxical demands.

One option involves "personalized informed consent," in which a medical practitioner allows the patient to decide whether they would like to hear about relatively rare risks, or whether they would prefer that the doctor withhold such information. This option still puts the patient in control of their treatment, and it may be more ethical than automatically providing information that could bring about a negative expectation effect.[38]

Each patient will have a different preference. Some might conclude that remaining in the dark gives them the best chance at maintaining the positive outlook that can, as we have seen, make all the difference. I suspect my own fears are often much worse than the truth, though, and so I would rather be told the relevant information so that my expectations are at least based on objective fact. Fortunately, for patients like me who prefer to know, it is possible to minimize the nocebo response by changing the way the information is presented, using a strategy known as "reframing." Abundant psychological research shows that people often respond very differently to the same piece of data depending on how it is phrased. Framing is already a well-known and well-studied tactic for advertising and marketing executives—it is the reason that foods are labeled "95 percent fat-free" rather than "5 percent fat," despite both phrases conveying the same thing. And it looks likely that the same technique could be employed to reduce nocebo side effects.

Consider a study from Australia's University of New South Wales, in which students believed they were signing up for a trial of a benzodiazepine medication, used to treat anxiety. In reality, they all received a dummy pill that would have no direct chemical effect on the body. In accordance with the standard procedure, the students were told about some of the expected benefits, such as muscle relaxation and lower heart rate, as well as the potential side effects, which included headaches, nausea, dizziness, and drowsiness.

In some cases, the information was framed negatively, with the emphasis on the number of people who would experience discomfort, such as the following:

> Possible side effects include drowsiness. Approximately 27 out of 100 people will experience drowsiness.

In other cases, the information was framed more positively, with an emphasis on the number of people who would remain free of side effects, such as the following:

> Possible side effects include drowsiness. However, 73 out of 100 people will *not* experience drowsiness.

Although the two sentences convey the same statistic by different means, the people in the positive framing group reported fewer short-term side effects upon ingesting the pill.[39] Whenever we are presented with this kind of information as patients, we should question whether it could be reframed more positively. Worst-case thinking doesn't prepare you—it promotes the worst case.

Just as important, we can learn to reappraise the symptoms that we *do* experience. Remember that nocebo responses can exacerbate side effects that originate from the direct action of a drug. In this case there's no point pretending that the discomfort doesn't exist, but medical practitioners can nevertheless change the way the patient interprets their experiences and the meanings they assign to them so as to minimize discomfort in the long run. The consequences for the patient's well-being can be profound.

In an extraordinary experiment of this kind, researchers at Stanford University's Mind & Body Lab helped treat a group of children and adolescents with severe peanut allergy. The patients were all undergoing "oral immunotherapy," which involved gradually exposing the body to larger doses of peanut protein over a six-month period. All being well, the patient should become less and less sensitive to the allergen, until they can eventually eat a whole peanut without a severe reaction; but the therapy itself can sometimes result in unpleasant feelings, such as hives, an itchy mouth, nasal congestion, and stomach pain. Besides being a source of immediate discomfort, these side effects often *feel* like the start of a full-blown allergic reaction, leading to increased anxiety about the treatment and a relatively high dropout rate. In reality, the side effects tend to remain rather mild—and rather than being the start of a dangerous overreaction, they can be seen as a sign that the immune system is responding to the stimulus, which is an essential step in the process of desensitization.

Could knowledge of this fact change patients' attitudes to their side effects? the researchers wondered. And might those altered attitudes then change their experience of the overall treatment? To find out, they designed an information program that aimed to change the patients' mindsets throughout the treatment by giving them written pamphlets and holding lengthy discussions with a trained health professional. In these sessions the researchers compared the side effects to an athlete's sore muscles after training—an uncomfortable feeling that nevertheless signals the building of internal strength. Along the way, the patients were given exercises that would reinforce their understanding, such as writing a letter to their future self, reminding themselves of the new ways to interpret their symptoms.

A control group went through similar meetings that focused more exclusively on ways to manage the side effects, such as taking the protein dose on a full stomach, drinking water, or taking antihistamines. While these discussions contained much practical advice, they always framed the symptoms as an unfortunate consequence that had to be endured, rather than a positive signal that the treatment was working. For safety, both groups were also taught how to identify any life-threatening symptoms, and experts were always available to discuss any serious concerns. (If you suffer from an allergy yourself, please do not try to create your own immunotherapy without medical supervision.)

The effect on the patients' feelings of anxiety was substantial, with the positive reframing significantly reducing their worries about the treatment. Importantly, this positive mindset reduced the reporting of actual symptoms as the patients progressed to larger doses of the allergen and, eventually, to real peanuts. The benefits of reframing were evident in the patients' subjective experiences as well as in biological measures of the treatment's success.

At the start and end of the therapy, the patients were given blood tests that allowed the researchers to detect an antibody, called IgG4, that is produced by the body in response to the ingestion of peanut protein. If it is present at suitable levels, IgG4 seems to inhibit other immune responses that would lead to a full-scale allergic reaction.[40] In the beginning of the study, both groups showed very little IgG4 on blood tests. By the end, however, the children and adolescents in the positive intervention group had ramped up production to a much higher level than that of the control group, reducing the symptoms that they experienced as the trial progressed.

Like all expectation effects, the change brought about by the altered beliefs can be explained by recognized physiological mechanisms. Chronic worrying can trigger low-level inflammation, which appears to disrupt the immune system's capacity to adapt, for example. Once they were primed with positive information, the participants in the positive intervention group may have been freed from this biological barrier, allowing their bodies to respond more effectively to the increasing doses of the peanut protein.[41]

Besides being a powerful example of the mind-body connection in action, the peanut allergy study also provides us with a perfect demonstration of a process known as "reappraisal," in which we look for positive interpretations of negative events. And as we shall now see, we can apply this technique ourselves whenever we are injured or ill.

## THE PAIN-RELIEF MINDSET

Let's begin by assessing how you currently think about pain or discomfort. Imagine you are suffering from a migraine or bad back, or that you have broken your arm. If you are like me, you may automatically fall into a trap of "catastrophizing" thoughts, in which the appearance of a symptom leads to the expectation that worse will follow.

Psychologists measure catastrophizing by asking patients to rate, on a scale of 0 (never) to 4 (always), statements such as these:

*When I feel in pain . . .*

- *I worry all the time about whether the pain will end.*
- *It's awful and I think it will never get better.*
- *I become afraid the pain will get worse.*
- *I can't seem to keep my pain out of my mind.*
- *I keep thinking of other painful events.*
- *I wonder whether something serious may happen.*

Each statement reflects a different type of catastrophic thinking; together they create a kind of self-perpetuating nocebo response.[42] The scale is a good predictor of the discomfort that people will experience following an operation, for example, and how long they will have to stay in the hospital.[43] The tendency to catastrophize also seems to contribute to the severity of migraines and headaches and to the symptoms of people suffering from chronic joint and muscle pain.[44]

Given what we know about the role of expectation in pain, the scientists Luana Colloca and Beth Darnall go so far as to suggest that catastrophic thinking is "like picking up a can of gasoline and pouring it on a fire."[45] It's taking the evolved responses to injury—which could serve as a useful warning in times of danger—and then amplifying them beyond any possible benefit.

The adoption of a "pain-relief mindset" can help to break this vicious cycle. Patients can be taught, for instance, about the nature of pain, including the psychological processes that can exacerbate our discomfort and the fact that our mental state can powerfully influence the symptoms.[46] Once they have learned to identify the beginning of catastrophic thinking, they are taught to reconsider the factual basis of their anxieties. While pain can be a signal of danger, for instance, the intensity of the sensation doesn't necessarily reflect actual tissue damage. (A migraine can be agonizing, for example, but it is very rarely the result of a serious neurological problem.) Similarly, if it feels like the pain is never-ending, it might help to remind yourself that you have overcome previous episodes. And if you have come to associate certain triggers, such as an important work meeting, with a flare-up, it could be worth asking whether the connection is really as inevitable as you assume.[47]

Each person may catastrophize in their own unique way, but as a general principle you can ask yourself the following questions whenever you notice yourself ruminating about your health: "Is this thought negative and alarming, positive and comforting, or neutral? What is the evidence for and against this thought? Is there a more palatable way to think about this?"[48] Finally, you may try to remember a few reassuring phrases, such as "my pain is in my brain," and "the sensations are real, but temporary," that can counteract more general anxieties and that emphasize the power of the brain's capacity to bring its own relief.[49]

Like any skill, reappraisal takes practice, but many studies have shown impressive benefits for the patients who learn to apply it. More than half of people with chronic pain report at least a 30 percent reduction in their symptoms when using this technique, with many patients experiencing as much as 70 percent improvement; reappraisal also reduces the number of days that migraine sufferers lose to their headaches.[50] These techniques could also help to relieve momentary discomfort, such as if you have burned yourself on your oven.[51] Strikingly, these psychological therapies lead to some lasting changes in the brain, including a reduction in the size of regions that are thought to handle catastrophic thinking. It is as if reappraisers have turned off the pain amplifiers.

While most of the research in this domain so far has focused on pain disorders, it's likely that this technique could offer relief for other unpleasant conditions. Catastrophizing is thought to worsen the symptoms of asthma, which could feasibly respond to reappraisal in the same way—if you remind yourself, for example, that your body is making sure you have enough oxygen.[52] There are also some signs that reappraisal reduces the severity of the common cold.[53] If you recognize your symptoms as a sign that your body is appropriately fighting the virus, you may reduce your discomfort.

By soothing the anxiety around long-term illness and alleviating doom-laden thoughts, reappraisal may even benefit the health of your heart. One study found that cognitive behavioral therapy (which includes sessions on the best ways to reduce catastrophic thinking) after heart failure successfully reduced the risk of developing further disease.[54] We will need many more studies on large numbers of patients to confirm that finding and to refine these therapies, of course. But the influence of nocebo responses on our health is undeniable. And crucially, these negative expectation effects can be neutralized.

This understanding could not be more urgent, since there is now strong evidence that some nocebo responses can be contagious. As we shall discover in the next chapter, the spread of negative expectations among people has contributed to many modern health scares. Contrary to the conclusions of the anthropologists and historians studying "self-willed death," human beings in developed countries may be *more* susceptible to suggestion than ever before—and we need every possible tool to counteract this very modern curse.

### HOW TO THINK ABOUT . . . PAIN AND DISCOMFORT

- When you are warned about the potential side effects of drugs, try to find out whether the same symptoms were also observed in the placebo group in the drug trial. (Your doctor may be able to provide that information, or you can often find such statistics on government websites such as www.CDC.gov.) If so, there's a good chance the side effects are the result of a nocebo response.
- Look more critically at the data representing the risks of side effects, and practice reframing. If you are told there is a 10 percent chance of developing a side effect, for example, try to focus on the fact that 90 percent of patients will remain free of that symptom.
- If you do experience a side effect, try to ask whether it might be a sign of the drug's healing action. Doing so will not only neutralize your anxiety but may actually improve the benefits of the treatment.

- Assess whether you are prone to "pain catastrophizing" using the scale on page 74. If so, try to notice when you start to ruminate on your symptoms. Awareness is the first step to breaking the vicious cycle.
- When you find yourself falling into catastrophic thinking, ask whether there is a good factual basis for your thoughts; if not, look for a way to reinterpret the situation more positively.
- Remember your understanding of the nocebo response and reinforce that knowledge whenever you can. Some studies have found that it helps to write a letter to yourself describing what you know; others suggest creating a social media post to share your thoughts.

# 4

# THE ORIGINS OF MASS HYSTERIA

## How expectations spread within groups

In May 2006 Portugal was beset by mysterious outbreaks of illness. The disease appeared to afflict only teenagers, who experienced dizziness, breathing difficulties, and skin rashes. Within a few days around three hundred students across the country had been affected. A virus or some kind of poisoning seemed the most likely pathogens, according to some experts; others believed it might be an allergic reaction to a certain kind of caterpillar or to dust in classrooms. None of the explanations seemed convincing, however. As one health expert remarked: "I don't know any agent that is so selective it only attacks children."

Inquiries eventually revealed that a popular teen soap, *Morangos com Açúcar* (Strawberries with sugar), was to blame. In the days before the first reported cases, the main characters on the show had been infected by a life-threatening virus that had led to very similar symptoms. Somehow the "virus" had jumped from the small screen to a handful of viewers, creating real physical symptoms—despite the fact that the illness in the program was totally fictional. Those children had then passed it on to their classmates, leading the cases to multiply. Portuguese adults were unlikely to have been dedicated viewers of the melodrama—and were less embedded in the teens' social networks—so they were less likely to develop the disease.[1]

Scientists call this kind of outbreak, with no physical vector, a "mass psychogenic illness." (Whereas "psychosomatic" can refer to our mental state exacerbating existing symptoms, "psychogenic" means that the origin is purely psychological.) Other notable cases range from mysterious dancing manias in the Middle Ages to the emergence of strange, uncontrollable facial tics among YouTube users.[2] The experiences are very distressing for those concerned, yet commentators in the past have often dismissed these conditions as "imaginary," willfully deceptive, or the result of a mental weakness that was of little relevance to most "normal" people. Much like the "voodoo deaths," mass psychogenic illnesses were considered to be rare occurrences that happen to other people.

The *Morangos com Açúcar* outbreak shows us just how easily psychogenic symptoms can be triggered in otherwise healthy populations. In this case, the cause was soon established, and the teens recovered, but cutting-edge research suggests that the same process of social contagion is helping to spread and amplify nocebo effects to millions of people. And it's not just suggestible adolescents who are affected; the research shows that anyone can be susceptible to the social transmission of psychogenic illness. Indeed, there is a good chance that you have "caught" a nocebo effect yourself without even realizing it—and it is only by learning to recognize the signs that you can protect yourself from being "infected" again.

## MIRROR, MIRROR IN THE BRAIN

To understand the ways that a nocebo effect could spread from person to person, we must first examine the origins of social contagion more generally. This arises through an essential component of the prediction machine, named the "mirror system," which allows us to build others' physical and mental states into our simulations of the world.[3]

The story begins with a monkey and some peanuts at the University of Parma in Italy. In the early 1990s Giacomo Rizzolatti's team of neuroscientists had been examining the neuronal activity that leads to purposeful movements—the messaging that tells your hand to pick up an ice cream cone, for example. To do so, they attached a sensor to a macaque's brain and recorded the electrical activity of its neurons as it grasped for a toy or brought a piece of food to its mouth. Over many trials, the researchers found that distinct groups of brain cells lit up for each action, with a separate pattern apparently representing each of the different intentions. As an important step to deciphering the brain's "neural code," this was a significant finding in its own right.

Just by chance, however, they found that the monkey's brain would also burst into action as it watched the researchers grasping its peanuts or toys—even though the monkey had remained motionless. Even more striking, the readings showed that the pattern of activity was extraordinarily similar to what happened in the monkey's brain when it was grasping the objects itself.[4] The brain appeared to be reflecting what it saw and then re-creating the experience itself, leading the team to describe the cells as *neuroni specchio*, "mirror neurons." These cells, they claimed, allow us to understand *immediately* what another person is doing without having to think about it consciously.[5]

Later research, in monkeys and humans, revealed that the brain's mirror system responds to feelings as well as actions. When we see another person expressing an emotion, we show heightened activity in the brain regions involved in emotional processing and in the regions involved in the display of those feelings—as if we were experiencing them ourselves.

Importantly, this internal mirroring can then lead to overt physical mimicry.[6] Recordings of electrical activity through the skin show that your cheek muscles begin to twitch very slightly when you see someone else smile; if they frown, the muscles in your eyebrows start to contract; and if they have screwed their mouth in a look of disgust or pain, you can't help but cringe a little—all because of the automatic activity of this mirror system. The tone and the speed of our speech will also shift toward our conversation partner's voice. Even our pupils tend to dilate or contract to match the eyes of the person we are viewing.[7]

Without our even noticing it, the presence of another person can therefore change our body as well as our mind. And these bodily effects apparently have a purpose—they increase our understanding of what the other person is feeling.[8] In one ingenious demonstration of this idea, researchers recruited cosmetic surgery patients undergoing botox injections, which temporarily paralyze the muscles of the face, and they asked them to describe the feelings people were displaying in various photographs. The botox patients found it much harder to recognize the emotions, compared with participants who had a "skin filler" injection that did not interfere with the facial muscles. The participants needed the physical mirroring to fully appreciate what the people in the pictures were feeling; without it, their emotional processing was disrupted.[9]

Humans don't just communicate with facial expressions, of course; we have words and symbols, which can also stimulate the brain's mirror system. If you hear the word "smile," you'll experience a trace of activity in emotional processing areas and may even experience small movements in the facial muscles, as if you were actually on the verge of breaking into a grin. Like our direct mimicry of other people's faces, this leads us to feel a shadow of the effect ourselves, despite there being no objective reason for feeling happier.[10]

Quite by chance Rizzolatti's team—and their monkey—had therefore uncovered a neural basis for empathy, explaining how our feelings can subtly pass from person to person through a kind of contagion. "When people use the expression 'I feel your pain' to indicate both comprehension and empathy, they may not realize just how literally true their statement could be," they later wrote.[11]

Most of the time we experience only a weak reflection of another person's feelings, of course. We don't feel full-blown joy whenever we see a picture of a lottery winner or experience extreme anguish whenever we see someone crying; their expressions are only going to modify what we're already feeling. But even small effects can add up if we spend a long time in someone's company, or if we have multiple interactions with different people who are all showing similar emotional profiles.

As an illustration of how far someone's feelings can spread, imagine that you become friends with someone with an astonishingly positive attitude, who is incredibly satisfied with her life. You might feel a bit glad for her, but could her joy really bring lasting happiness to your life, too? According to one detailed longitudinal survey—the Framingham Heart Study—the answer is yes. Because of your regular interactions with her, you would be 15 percent more likely to achieve a high score on the survey's measure of life satisfaction—despite no direct change in your immediate circumstances.

How about your friend's friend? The same study found that their happiness will be passed on to your friend, who passes it on to you, increasing your chances of happiness by about 10 percent in the coming months. Your current satisfaction with life is even influenced by a friend of a friend of a friend, who can increase your chance of happiness by about 6 percent. These are people you almost certainly have never even met, and you probably don't even know of their existence, but they are nevertheless influencing your well-being through a chain of interactions.[12]

The discoveries of the mirror system, and the extent of social contagion more generally, have important consequences for our mental health, revealing just how much our well-being depends on the concentric rings of our social circle. But they can also shed light on the ways that symptoms spread through a group during mass psychogenic illnesses. When we are in a group of people who are all extremely concerned about the threat of a biological weapon, for example, each person can begin to amplify the others' fears—creating a kind of echo chamber that puts everyone into a state of panic.[13] Even more important, our overactive empathic brain might then begin to simulate feelings such as pain, nausea, or vertigo that another person is reporting. If we are lucky, this effect may not be strong enough to have a real impact on our well-being. But if we are already in a situation where illness seems likely, the simulations of the mirror system could feed into the prediction machine's calculations, creating or exaggerating a nocebo response. And the more we interact with people who are unwell, see their suffering, and talk about their symptoms, the worse we will feel.

Giuliana Mazzoni, a psychologist at the University of Hull in the UK, was one of the first to reveal how potent this process could be. She invited a small group of participants to take part in a "study of the individual reactions to environmental substances." In pairs, the subjects were asked to inhale a suspected toxin that had been reported to produce headache, nausea, itchy skin, and drowsiness. (In reality, it was just clean air.) Unbeknownst to the participants, however, the "partners" in the trial were really actors who had been told to deliberately feign the symptoms as they inhaled the gas. The consequences of this observation were startling. People who saw a partner in discomfort reported much severer symptoms themselves, compared with people who did not see the reported side effects.[14]

Mazzoni's results were first published in the late 2000s, and we now have a plethora of other studies showing that nocebo-like symptoms can jump from person to person through social contagion. One study mimicking a drug trial found that participants taking an innocuous pill reported *eleven times* as many symptoms—such as nausea, dizziness, and headaches—after observing an undercover actor feigning sickness.[15] Another examined repeat visitors to blood donation clinics. It is not unusual for people to feel faint or dizzy after their pint has been taken, but these symptoms were twice as likely to occur if the donor had just seen another visitor on the verge of collapse.[16]

These socially contagious effects are highly specific: it's the particular symptoms of the other person that are transmitted and heightened during observation, rather than general feelings of unwellness. And they occur over and above the typical nocebo response you might get from a written or spoken warning from someone who is not also displaying the symptoms.[17]

Tellingly, your susceptibility to these effects seems to reflect your overall capacity for empathy and your ability to dampen those feelings when necessary. One standard measure of empathy asks people to rate statements such as "I am often quite touched by things that I see happen," "When I watch a good movie, I can very easily put myself in the place of a leading character," and "When I see someone who badly needs help in an emergency, I go to pieces." Perhaps because they have a more reactive mirror system, people who score high on these kinds of questions are more likely to absorb other people's signs of illness and report the same symptoms themselves; they are also more likely to feel better if someone else shows relief.[18]

The most striking evidence of expectation contagion comes from Fabrizio Benedetti, a neuroscientist at Italy's University of Turin. Benedetti has been at the heart of research into the placebo and nocebo effects and their roles in our health. But he also happens to investigate the effects of altitude on fitness at the Plateau Rosa research station, which lies on a snow-clad mountain in the northwestern Alps, at around 11,400 feet above sea level. The location—which remains open for skiing throughout the year—provided an ideal setting to test the way that expectations of illness can spread throughout a group in a nonclinical setting.

The study in question focused on the "altitude headache" reported by many climbers and skiers, which is thought to be a direct effect of thinning air at great heights. There's little doubt physiology does play a direct role in this phenomenon: to cope with low oxygen, our blood vessels dilate, for example, which is thought to increase the pressure in the brain's capillaries. A nocebo response should vastly increase the discomfort, and Benedetti wanted to investigate whether social contagion could spread and amplify that negative expectation effect between people. To do so, he invited 121 students from the local medical and nursing schools to take a three-hour journey up to his high-altitude lab—reachable through three successive cable cars. The students were all attending the same course and knew one another. Rather than warning each of them individually of the potential effects of altitude, Benedetti's team selected just one student—the "trigger"—to be primed with the expectation of a headache. This trigger was shown a flyer that explained the risk, and a video of a sufferer in bed and grimacing with pain (the kind of scene that is likely to spark empathic feelings). Afterward, he or she was asked to call the researchers two days before the trip to confirm the right dose of aspirin to bring along.

Benedetti's team did not tell the student to pass on the message, but the trigger happened to mention it to a few friends anyway, who then mentioned it to their own acquaintances. By the time of the actual trip, news of the potential risk had reached 35 others, all of whom had called the center asking for advice about the amount of aspirin to take with them.

The effects on their health during the course of the visit were staggering. When Benedetti surveyed the group, 86 percent of these primed students experienced an altitude headache, compared with 53 percent of the students who had not heard of the risk from their classmate. What's more, the average intensity of the headaches was also much greater for the people who had been in touch with the trigger. Using saliva samples taken after they had reached the high-altitude lab, Benedetti found that these differences were even reflected in the participants' brain chemistry, which showed an exaggerated response to many of the known changes that occur in response to thinning air. The participants who had spoken to the trigger showed higher levels of prostaglandin molecules, for instance, which are thought to lie behind the vasodilation that may cause altitude headaches.

In a stroke of genius, Benedetti asked the students to report how they'd heard about the headache, and who they'd discussed it with since—allowing him to map the spread of contagion through the group. He found that the more times they had discussed the symptoms, the worse their headaches, and the higher the levels of prostaglandins. Each interaction had amplified their anxiety and made them feel more pain as a result, with real changes in their neural chemistry.[19] "It doesn't seem to matter where expectations come from. They could be generated by a doctor or a peer—but the stronger expectations are, the stronger the effects will be," Benedetti told me.

Perhaps counterintuitively, contagious expectation effects like this could serve a useful purpose—particularly when the risk of a physically transmitted disease is high. Imagine that you live in a region with dangerous ticks or malaria-carrying mosquitoes. If you see people scratching around you or hear them talking about an itch, your brain can heighten your skin's sensitivity so that you are more likely to detect the presence of the insects and remove them before they transmit disease. Similarly, if you've been eating in a group and one of you falls ill, "catching" their nausea might feel unpleasant, but it could prevent you from continuing to consume a potentially dangerous pathogen. Humans are social animals, after all, and the prediction machine is simply using all the cues it can to prepare you for potential illness or injury.

Most of the time, this process functions perfectly well. In certain circumstances, however, it can spark large outbreaks of sickness that have absolutely no physical origin.

## THE THREE LAWS OF CONTAGION

Now that we understand the mechanics of how individuals mirror each other's physical symptoms, we can resolve many medical mysteries—and identify the precise conditions that make a mass psychogenic illness more likely.

If we return to Portugal in 2006, it is easy to imagine how the viewers of *Morangos com Açúcar* may have felt so involved in the drama, with their brain's mirror system re-creating the characters' feelings of illness. Once a few of the teens had started to demonstrate physical symptoms, their displays of illness may have come to infect their classmates' minds, who then infected others, in some cases leading to hospitalizations and school closures. After the authorities started naming the plausible causes—like the presence of poisonous dust or the presence of dangerous caterpillars—the cases would have only multiplied, until they eventually announced the real psychogenic origin.

Similar processes could lie behind many reports of mass psychogenic illness throughout history. In the prescientific age these outbreaks took the form of violent convulsions, fainting fits, and even a series of dancing manias that afflicted whole towns and villages in medieval and early modern Europe. Across the Atlantic, an outbreak of mass psychogenic illness may have led to the Salem witch trials of 1692. The reports of paranormal possession began with two cousins—Betty Parris and Abigail Williams—who suffered from convulsive seizures, and within days the symptoms had spread to the other adolescent girls in the town. Some modern doctors have argued that the fits may have been caused by ergotism—poisoning by a fungus that had infected the residents' crops—but others suggest that the reports bear all the hallmarks of a mass psychogenic illness. It's possible that one could lead to the other, of course—perhaps Parris or Williams had some kind of organic illness, but the symptoms then spread to others through a contagious expectation effect.[20]

In the nineteenth and twentieth centuries, these episodes became much rarer. Mass psychogenic illnesses more regularly took the form of apparent poisonings that turned out to have no physical origin. One of the most noteworthy concerned the workers at a mill in Spartanburg, South Carolina, who in 1962 began to experience nausea, cramps, bodily weakness, dizziness, and extreme fatigue. Rumors soon began circulating that a poisonous insect had arrived in a shipment of textiles from England. Within weeks, around sixty workers had fallen ill. Experts from the Communicable Disease Center (now the Centers for Disease Control and Prevention) combed the entire plant for the culprit. They found black ants, houseflies, gnats, beetles, and mites—but none of these species could have caused the illness. In the end, they traced the outbreak to a twenty-two-year-old mill worker, who told her friend that she thought she'd been bitten, and who subsequently fainted. All the other cases then arose through social contagion.

Interviewing the workers, sociologists found that two factors could predict which particular employees would become victims of illness. The first was the amount of stress they had been experiencing recently: employees with marital difficulties or family problems were more likely to be affected than those with stable circumstances. The second was their connection to the other victims: if they personally knew another sufferer and had interacted with them regularly, they were more likely to be struck down.[21]

These could be considered the first two laws of contagion. The third—and last—concerns the environment: whether there is a feasible threat that would raise overall expectations of illness. It seems unlikely that the mill workers had been worried for long about the risk of English insects, but in some climates, fears of impending illness are never far from mind, making the spread of symptoms far more likely.

This might explain why mass psychogenic illnesses appear to be especially common during times of political upheaval or warfare. In 1983, for example, students and staff at a Palestinian girls' school in the West Bank began to experience blurred vision and breathing problems, accompanied by the smell of a rotten egg. As news of the outbreak spread, nearly a thousand students in the area fell ill. Eventually, epidemiologists managed to trace the outbreak at the original school to a broken latrine emitting an unpleasant odor, which the girls had interpreted as a poisonous gas. The closer the classrooms were to the toilet, the more likely students were to have reported symptoms on the first day. During their breaktime, those girls discussed the danger with their friends, who discussed it with their friends—just like the students in Benedetti's altitude sickness study. Soon the expectation of illness spread across the whole school, and—as the reports became better known—to other establishments across the region.[22]

The United States suffered similar outbreaks following the 9/11 attacks. In late 2001 and 2002 there were widespread fears of further Islamist attacks, including the possibility of biological warfare. It began with reports of rashes in Indiana and spread to northern Virginia, then Pennsylvania, Oregon, and Massachusetts. Strangely, the symptoms seemed to be limited to specific locations: the rashes would become angrier in school but slowly calm down when children returned home. Needless to say, the outbreak caused tremendous anxiety among parents, but scientists failed to identify a weapon or any other environmental cause. They considered the use of pesticides, mold in the building, and even an allergic reaction to chemicals used to manufacture textbooks—but none of the explanations stood up to careful scrutiny.[23]

An expectation effect transmitted through social contagion may even shed light on the "Havana syndrome" first experienced by US diplomats and intelligence agents posted to Cuba. In the last days of 2016, a CIA operative in Havana arrived at the US embassy reporting strange symptoms: dizziness, ear pain, tinnitus, and a fuzzy head. The strangest element was the apparent source of his discomfort: at home, he had had the distinct impression that an intensely irritating loud noise was following him from room to room. The sound went away, he said, only when he opened his front door.

As news of his experience spread, more colleagues came forward to say that they too had noted the same bizarre symptoms over the previous months. The descriptions of the sound varied from extremely high-pitched sounds ("a teapot on steroids") to a "baffling sensation akin to driving with the windows partially open in a car."[24] Some reported feeling shaken by a vibration or "pressure" that woke them up at night; others heard nothing at all but still felt a sense of disorientation, confusion, and vertigo. What was clear was that the sensation was highly unpleasant and accompanied by concussion-like symptoms, leading the US government to declare that an acoustic weapon was being used to intimidate its diplomats and intelligence agents.

The fears soon spread to other countries' diplomats, with Canadian staff reporting very similar symptoms, as well as nose bleeds and insomnia. Then came apparent attacks of a similar nature from other countries thousands of miles from Havana: the State Department evacuated staff from its embassy in Beijing and consulates in Shanghai and Guangzhou after an outbreak.

Acoustic scientists struggled to identify any possible way that a sound wave could be directed, at a distance and at sufficient intensity, to cause significant damage to the human brain. Indeed, an analysis of a supposed recording of the sounds tormenting the embassy staff revealed that the waves were the sound of cicadas. There is still a lively scientific debate about the ultimate cause of these symptoms; some scientists have argued that it may have been the result of a weapon emitting focused radio waves. Others, however, are convinced that the illness was psychogenic. The signs of Havana syndrome certainly bear an uncanny resemblance to many of the other symptoms that are known to arise from noxious expectations. And the tight-knit community of expats living under high stress in a foreign country would have provided the perfect social environment for psychogenic illness to break out and spread from person to person.

As we also saw with the Salem witch trials, the potential presence of psychogenic illness does not preclude an environmental origin. Some—currently unknown—physical cause may have triggered illness in a small group of people, the symptoms of which then spread through observation and expectation to many others who had no direct contact with the original threat.[25]

Most interesting, from my perspective, were people's reactions to this possibility—and how little they seemed to understand the power of our expectations to create illness. "To artificially display all of these symptoms, you'd have to actually go and research, practice, be the most consummate actor ever, and convince one expert after another," one of the doctors who had been involved in the initial diagnosis said at the time.[26] Senator Marco Rubio, who chaired a special hearing on the attacks, took a similar line, describing mass psychogenic illness as "a bunch of people [who] are just being hypochondriacs and making it up."[27] As the abundant scientific research on expectation effects demonstrates, this could not be further from the truth. There's nothing artificial or fantastical about mass psychogenic illness—it is a natural consequence of our socially sensitive minds and the prediction machine's astonishing capacity to preempt threats.

## VIRAL THOUGHTS

While the mass psychogenic illnesses discussed so far were extremely troubling events, they affected only a limited number of people in isolated communities. And once the physical risks had been eliminated, many of the patients' symptoms began to dwindle. Other outbreaks have not been so easy to quell, however—thanks in a large part to traditional and social media.

These health scares may start as regular nocebo effects—caused by a fear of the unknown or even a reasonable warning from a health professional—which are then shared with near acquaintances. Once the cases have reached a critical mass, documentaries, online articles, and social media posts will then spread the news far and wide—often with highly emotive first-person accounts that kick the brain's mirror system into action. This in turn will lead many more people to develop symptoms, and over a short period of time the symptoms can start to affect thousands or even millions of people.

Let's first consider the "technopanic" caused by the introduction of new technologies. People have often been fearful of innovation, setting up an anxiety that can promote the transmission of a noxious expectation effect. This results in the spread of symptoms, first through direct social interactions and then through media coverage. As early as 1889 the *British Medical Journal* reported a sharp uptick in cases of "aural overpressure," resulting in a continued buzzing in the ear, "giddiness," "nervous excitability," and "neuralgic pains." The culprit? Alexander Graham Bell's newfangled telephone.[28] Similar outbreaks have accompanied the rise of the telegraph, the radio, and computer monitors—devices that few today would consider to be serious health risks.[29]

More recently, the rise of wireless technology has resulted in reports of headaches, breathlessness, insomnia, fatigue, tinnitus, dry eyes, and memory problems in the presence of Wi-Fi or 5G signals. While this may seem to be a niche concern, "electrosensitivity" has affected a sizable number of people—from 1.5 percent in Sweden (around 150,000 people) to 4 percent in the UK (around 2.6 million people).[30] The sufferers believe that long-term exposure to electromagnetic fields can disrupt the signaling between neurons and may lead to cellular damage in the long term. Yet laboratory studies show that the low doses experienced in our homes or offices are certainly not powerful enough to cause any harm.

To find out whether a psychogenic origin might instead be the explanation, James Rubin at King's College London invited sixty "electrosensitive" people into his lab. He gave each participant a headband with a mobile phone antenna above one ear. In some trials it emitted a signal; in others it did not. For a period of fifty minutes, the participants were asked to note any of the symptoms they experienced. If electrosensitivity arises from the physical effects of the electromagnetic field, you would expect far more symptoms to have been reported during the real exposure, compared to the sham trials. In reality, the subjects in the control condition were actually a little *more* likely to report problems such as headache (despite the fact that no electromagnetic waves were being emitted). That seriously undermines the notion that side effects are caused by an inherent biological reaction to the electromagnetism.[31] "I've got no doubt in my mind that people are genuinely experiencing physical symptoms," Rubin told me. But those symptoms are the result of expectation and social contagion, not radiation.

Rubin's study was published in 2006, and later experiments have shown that healthy people who have never experienced electrosensitivity before are much more likely to report symptoms after seeing an alarmist video outlining the "dangers."[32] Importantly, this online information often includes clips of individuals directly sharing reports of their illness—and as we have seen before, seeing and hearing someone with symptoms makes it so much more likely that the nocebo effect will catch on.

Equally common—but much more problematic for global health—are psychogenic reactions to vaccinations. Many people receiving the influenza vaccination, for example, report that they have suffered symptoms such as fever, headaches, and muscle pain; some even claim that you can catch full-blown flu straight after receiving the vaccine. (According to one recent survey, around 43 percent of US citizens now believe this to be the case.[33])

The truth here is a little more complicated than in the reports of electrosensitivity. There are various forms of flu vaccine, but any flu vaccine delivered by injection contains either an inactivated form of the virus or a single protein taken from the virus. In both cases, the altered virus or its protein is unable to replicate within the body—meaning that the vaccine cannot lead to infection. It's impossible. For injected vaccinations, clinical trials show that people receiving a placebo are just as likely to exhibit the symptoms of flu as are the patients receiving the actual vaccine.[34] According to the US Centers for Disease Control and Prevention, the only difference is that the real vaccine is slightly more likely to make your arm feel sore at the site of the injection.[35]

The case of certain flu vaccines delivered through a nasal spray is more complicated. These contain an "attenuated" virus, which has been weakened but is still potentially active. The virulence of the virus has been reduced so that it cannot lead to full-blown flu, but there is some evidence that it can lead to mild symptoms, such as a runny nose and moderate fever over the following days. Even so, trials suggest that this direct biological action can explain only a minority of the reports, and that many people's symptoms—particularly the sensations of fatigue or headache—may be psychogenic.[36]

In either case, doctors' warnings might have given rise to some of the symptoms—but the chances of experiencing them will be much greater if you know of a relative or friend who has also felt that discomfort, or if you have seen social media posts complaining of the side effects. And the consequences of that social contagion can sometimes be dramatic. During the 2009 swine flu outbreak, forty-six Taiwanese middle school students were taken to the hospital with severe sickness following vaccination, yet doctors found that their symptoms were purely psychogenic.[37]

Mass psychogenic illnesses of this kind have disrupted many other inoculation programs, including an outbreak in Colombia in 2014 during the rollout of the HPV vaccine. It began with a few schoolgirls in Carmen de Bolívar, who reported severe sickness after receiving the vaccine. Soon videos of the unconscious and twitching girls were uploaded to YouTube and shared in the mass media—resulting in six hundred further cases within the following weeks.[38] Once again, investigations showed that the symptoms were purely psychogenic—but the event had disastrous consequences for the program, with a huge drop in the vaccine's uptake over the following years.

Very similar patterns can be seen in the side effects of statins. These drugs are widely prescribed to reduce blood cholesterol, which can clog the arteries and increase the risk of heart disease and stroke, and there is strong evidence that they can significantly improve a patient's longevity. In the early 2010s, however, patients started to raise concerns about side effects, including chronic muscle pain, that appeared to arise from the drugs.[39] These worries were covered by numerous media outlets, which interviewed patients describing their agony and published photos of people experiencing extreme pain—just the kind of content that is going to start activating the brain's mirror system.[40] As a result, thousands of people began to report symptoms and stopped taking their drugs.

Placebo-controlled trials, however, have shown that the rate of side effects among people taking statins is almost as high as that among people taking an inert pill.[41] (According to a review by the American Heart Association, the difference is less than 1 percent.[42]) Yet people's fears have been difficult to quell, and the rapid rise in cases confirms the ways that individual patient reports, amplified by the media and shared widely on social networks, can quickly create a mass psychogenic illness based on relatively rare events.

One comparison of thirteen different countries found that the accessibility of negative coverage online is directly proportional to the percentage of patients experiencing side effects in that region. In the United States and the UK—where negative stories about statins are most frequently encountered—the proportion of patients reporting muscle pain is about 10–12 percent, whereas in Sweden and Japan it hovers at around 2 percent, which is much closer to the rates predicted by placebo-controlled trials.[43]

Perhaps the most prevalent expectation effect concerns the rise of certain food intolerances, which are becoming increasingly common in Europe and the United States. Consider some of the digestive problems associated with gluten, the proteins found in wheat, rye, and barley. Around 1 percent of people are thought to suffer from celiac disease,[44] which is caused by an overreactive immune system that mistakes dietary gluten for a dangerous pathogen.[45] The resulting damage to the gut impairs the body's ability to absorb nutrients and can lead to anemia and other deficiencies. A further 1 percent of adults may be affected by wheat allergy, in which other proteins in the grain, besides gluten, trigger an exaggerated immune response, resulting in immediate symptoms such as vomiting and itchiness.[46]

A third group, however, report a less easily defined "gluten sensitivity." People with this condition don't show the damage to the gut seen in celiac disease, or the release of antibodies that characterizes wheat allergy, yet they still report abdominal pain, bloating, diarrhea, and headaches.[47] And the latest research suggests that people's expectations may often be responsible for this discomfort. In blind trials, for instance, suspected sufferers cut out all gluten from their diet for a few weeks before they were asked to eat products such as bread or muffins that may or may not contain the proteins. Combining the results of ten different studies, a recent meta-analysis found that 16 percent of people with reported gluten sensitivity actually responded to the gluten, but not to the placebo, while a much larger proportion—around 40 percent—responded equally to both groups of food, suggesting that expectation had played a large role in their symptoms.[48] (Importantly, many of these studies had excluded placebo foods that might have contained so-called FODMAP carbohydrates, which have also been touted as a potential cause of the symptoms.)

Each person's case will have to be individually assessed—but based on these results, a nocebo effect is a probable cause for a large number of sufferers. The glut of lifestyle magazines and websites describing the dangers of wheat—and the continued dinner party conversations that have arisen from this coverage—has quickly accelerated the spread of these negative expectations about the foods we eat. In the mid-2010s the number of people in the UK describing gluten sensitivity grew by 250 percent over three years, rising to around a third of the total population—an astonishing increase that is highly unlikely to arise from any physical source.[49] Data is sparse for other regions, but the trend does seem to be catching on in many other countries.[50]

## THE STENCH OF STIGMA

These examples are just some of the ways that expectation effects, spread or amplified through social contagion, are currently afflicting world health, but there are doubtless many more. Most recently, the physiological effects of noxious beliefs may explain some people's reactions to wearing masks during the Covid-19 pandemic—with a sizable number of people arguing that the face coverings left them short of breath and suffering from migraines. Most masks are made of relatively light fabric that should not have obstructed breathing, but the negative expectation of suffocation could have contributed to the appearance of these symptoms, which were then widely discussed on social media.

Having heard members of my family describe these experiences, I initially experienced some of the headaches and shortness of breath myself. My susceptibility isn't surprising; I score fairly high on the test of empathy (page 85) that is thought to reflect the reactivity of our mirror neurons. Thanks to my knowledge of the nocebo effect, however, I was able to question the origins of those symptoms, and I soon found an online video of a cardiologist performing a workout in a surgical mask—without demonstrating any loss in his blood oxygen level. The demonstration was enough to recalibrate my brain's predictions, and the uncomfortable feelings soon went away—providing me with another example of our power to reframe and reinterpret our feelings.

Anytime we have a new technology, medical procedure, or change in diet, the unfamiliarity of the innovation will create distrust and fear that could lead noxious expectations to infect a population. The challenge for health authorities is to distinguish between the actual physical risks and the product of expectations, and to address people's needs accordingly; to ignore either aspect will do the patients a great disservice. In many cases, people's symptoms do slowly decline once the possibility of a physical threat has been eliminated and their brain has updated its predictions—but that can only work if the patients trust the experts bringing the news. If experts don't deliver the message in a sensitive manner, patients are likely to discount the psychogenic explanation and may even decide that there is some kind of cover-up by the medical profession. This will not only exacerbate their own suffering, but also increase the chance that they will transmit the expectation of illness to other people.

This is why we urgently need greater public awareness of expectation effects more generally. Fortunately, there is now evidence that teaching people about nocebo effects, and their power, can help protect them from future mental pathogens. Keith Petrie and Fiona Crichton at the University of Auckland in New Zealand, for instance, have documented the rise of "wind turbine syndrome"—a mass psychogenic illness caused by a fear of the low "infrasound" waves created by the blades of wind turbines. The symptoms are incredibly unpleasant, including headache, earache, tinnitus, nausea, dizziness, heart palpitations, vibrations within the body, aching joints, blurred vision, upset stomach, and short-term memory problems—but careful studies show that they all arise from people's expectations and the social contagion of symptoms, whether or not infrasound waves are actually present.[51] Petrie and Crichton, however, found that a clear explanation of the nocebo response and the power of expectation to create physical symptoms could "immunize" people from the illness.[52] To prevent needless suffering, this kind of information should be incorporated into public health messaging about issues of emerging concern that may be psychogenic in nature.[53]

On an individual level, we can all try to be more discerning in the ways we think about potential new health scares. Be aware that personal stories, while compelling, do not provide strong evidence of a real danger, and that the symptoms people report may arise from many sources. Check that media reports are based on reliable scientific research and look for comparisons of symptoms in people who have and have not been exposed to the supposed threat. (Like the placebo-controlled drug trials, any good study should ideally include some kind of "sham exposure" that can test whether expectations have played a role.) If there is no difference between those populations, you probably don't need to worry: the symptoms are largely the product of expectations. Even if there is a difference, try to take note of whether the absolute risk is high or low. For many health scares—like statin intolerance—the purely biological side effects are still very rare. (If you have serious concerns for your health, you should of course consult a doctor.)

As people increasingly come to appreciate the power of expectation to create symptoms, we need to abandon the stigma attached to psychogenic and psychosomatic illnesses. Society has, after all, made important strides in advancing our conversations around mental illnesses such as depression and anxiety. Yet people are—quite inexplicably—much more dismissive of conditions that may arise in the mind but then influence the body. According to one of the specialists I spoke to, this stigma is sadly prevalent among medical professionals, who may communicate their disdain to their patients.

The truth is that we are all susceptible to expectation effects that can bring about real physical discomfort. Discovering that fact should be no more shameful than having a regular infection, a broken bone, or clinical depression. Psychogenic and psychosomatic symptoms are a natural result of the brain's extraordinary prediction machine, and a recognition of their psychological, social, *and* cultural origins will be essential as we move on to explore the consequences of our expectations for exercise, diet, stress, and sleep in the following chapters.

## HOW TO THINK ABOUT . . . HEALTH SCARES

- Be aware of the people around you and the ways that your body might start to mimic their mental and physical state through the brain's mirror neuron system.
- Bear in mind the particular situations that are likely to give rise to a mass psychogenic illness—such as times of high political anxiety, the introduction of new technologies, and the adoption of new medical practices. Try not to associate "unfamiliar" with "dangerous."
- When considering cases near to you, remember the potential role of coincidence. (A vaccine may have appeared to bring about an illness in your friend, for example, but they probably caught the infection before receiving the injection.)
- Apply critical thinking to the health news you read. Look for reliable scientific sources and try to find data on people who have and haven't been exposed to the alleged threat. Don't just rely on anecdote or personal stories, however convincing they may seem.
- If you feel sick with a potential psychogenic illness, seek medical advice, but be open-minded about the possibility that your symptoms could be the result of an expectation effect. Once your beliefs have become entrenched, it may be far harder to undo the effects.
- Avoid stigmatizing language when thinking about yourself or others. Stigma will only make it harder to question the beliefs that may be causing an illness or exacerbating its symptoms.

# 5

# FASTER, STRONGER, FITTER

## How to take the pain out of exercise

It is July 18, 1997, at stage 12 of the Tour de France, and Richard Virenque of the French Festina team is preparing for the individual time trial in Saint-Étienne. Virenque's specialism is the rugged mountain stages, not the time trials, but he has heard of a new drug that will deliver a spurt of energy for the fifty-five-kilometer circuit, and so he asks his physiotherapist, Willy Voet, to procure the "magic potion." The team are no strangers to performance-enhancing drugs, so Voet's initial objections are practical rather than moral; he fears trying a completely new substance in the middle of the tournament in case a bad reaction scuppers Virenque's chances. After some persuasion, however, he agrees to meet with the *soigneur* pushing the drug, and he is soon in possession of a small jar of a mysterious white liquid, which he is told to inject into Virenque's buttocks before the event.

On the day of the race Voet faithfully delivers an injection—and the results are breathtaking. Virenque races head-to-head with his great rival Jan Ullrich for much of the race. Although the German eventually wins with a time of 1 hour 16 minutes and 24 seconds, Virenque is 3 minutes and 4 seconds behind him—a much better result than he could have imagined. "God I felt good!" he later told Voet. "That stuff's just amazing." It was, Voet said, "the time trial of his life."

Little did Virenque know that there was no active ingredient in the magic potion. Before delivering the injection, Voet had swapped the mysterious white substance for a glucose solution. The confidence boost—combined with the cheering of the crowds—was all Virenque needed to perform at his best. This time, at least, he hadn't broken any rules.

"There is no substitute for self-belief," Voet would later write in his autobiography. "There was no more effective drug for Richard than the public. A few injections of *allez Richard* going around his veins, a big hit of adoration to raise his pain threshold, a course of worship to make him feel invincible. That was the sort of gear Richard needed."[1]

Stories of such dramatic performance boosts are commonplace in sports. You can train your body day in, day out for years on end—but ultimately it is your mindset that will decide your physical limits.

The middle- and long-distance runner Paavo Nurmi (1897–1973)—a nine-time Olympic gold medallist nicknamed the Flying Finn—expressed as much when he said, "Mind is everything; muscles, pieces of rubber. All that I am, I am because of my mind." So did Roger Bannister, the first man to break the four-minute mile, in 1954: "It is the brain which determines how hard the exercise systems can be pushed," he wrote in his autobiography.[2] It's also the philosophy of Eliud Kipchoge of Kenya, who is arguably the greatest marathon runner of all time. "I always say I don't run by my legs, but I run by heart and my mind," he explained. "What makes a person run more is their mind. If your mind is calm, and well concentrated, then the whole body is controlled."[3] At the time of writing, Kipchoge has won thirteen of the fifteen marathons he's entered—and holds the world record of 2 hours 1 minute and 39 seconds.[4]

Despite the prevalence of this idea in sports folklore, it has taken sports scientists a century to understand the true power of our minds to influence physical performance. Following the surge of interest in medical placebos, however, researchers are now enthusiastically investigating expectation effects in fitness and sports. At the heart of this is new work on the brain's role in regulating our energy expenditure and in creating the physical sensations of strain and fatigue. The prediction machine estimates how far it can push the body without causing damage, and when it believes it is reaching its limits, it puts the brakes on our performance, creating the sense that we are "hitting the wall"—whether that is in the middle of a five-kilometer run or the last stretch of an Ironman triathlon.

These findings may help professional athletes win world records, but they are even more relevant for reluctant exercisers who struggle to maintain a fitness regimen. By adopting the right mindset, even the most devoted couch potato can enjoy more gain, and less pain, from their workouts.

## MIND OVER MUSCLE

Much like the research on placebos and nocebos, our new understanding of exercise has come in fits and starts—beginning with the work of the Italian physiologist Angelo Mosso in the late nineteenth century. In painstaking experiments at the University of Turin, he attached small weights to people's middle fingers. The participants had to move the fingers until they reached exhaustion, while Mosso recorded the strength of the muscle contractions using an "ergograph." (Finger curls might seem a rather trivial exercise, but it was appealing for the experiment because Mosso could control and measure the movements so precisely.)

As you might expect, the participants started out strong, but the movements became more and more arduous over time as their muscles fatigued, and physical exercise undertaken beforehand meant that they experienced that fatigue after fewer movements. Importantly, however, Mosso found that purely intellectual tasks—such as lecturing or grading university examinations—could also lead to a more rapid decline in their muscular force. Based on this and many other experiments, he concluded that our sense of fatigue comes from two different sources—a "psychic process" from the exhaustion of the brain-based "will," and the buildup of chemical "poisons" in the muscles. "Fatigue of the brain reduces the strength of the muscles," he wrote in *La Fatica* (*Fatigue*), his great work on the subject. And if we want to increase endurance, we need to train the mind as well as the body, he said—the two are intimately connected.[5]

If the course of scientific history were just, Mosso would have been widely recognized for his work on physiology and neuroscience—and sports scientists would have continued to investigate the many psychological factors that influence our strength and endurance. But Mosso died in 1910, and later scientists focused almost exclusively on biochemical changes in the muscles. "He was essentially written out of history," Timothy Noakes, a sports physiologist at the University of Cape Town, told me.

According to the prevailing theory, our muscles tire when they run out of fuel in the form of the molecule glycogen, which is stored in the tissues, and with the buildup of toxic by-products such as lactic acid, which make it harder for the fibers to contract, slowing your movements. (Since lactic acid is also a product of fermentation, your muscles are essentially being "pickled," according to this theory.) This would be especially problematic with prolonged or intense exercise if our heart struggles to pump enough fuel and oxygen around our body to replenish the supplies and if our muscles are working so hard we don't leave ourselves enough time to convert the lactic acid back to glycogen.

Other factors—such as dehydration and body temperature—were also thought to play a role in setting our physical limits, but the mind was thought to be of much less importance. An athlete could try all they wanted to pace themselves to avoid using up all their energy too soon, but if they pushed too hard, they were going to "hit the wall," and their psychology could do little to recruit more muscle fibers or diminish the physical sense of exhaustion. If one athlete is better than another, it is simply because they are able to produce energy more efficiently with fewer toxic by-products, thanks to their training and to the luck of the genetic lottery.

This biochemical explanation of exhaustion has stuck for decades—you were probably taught it in biology classes at school. During the last few years, however, the foundations of the theory have started to crumble following a series of puzzling discoveries. Notably, scientists have tried and failed to find evidence that most athletes are performing at maximum capacity, as predicted by the biochemical theory. Rather than showing a plateau or decline at the point of exhaustion, for instance, athletes' heart output and oxygen consumption appear to be high enough to maintain their exercise for longer—yet they still hit the wall regardless.

Even more problematic for the accepted theory are the studies examining the activity of our muscles as they move. By attaching electrodes to subjects' arms and legs, researchers have found that only 50–60 percent of muscle fibers appear to be operating during prolonged or intense exercise. If biochemical changes in muscle fibers were the sole cause of physical fatigue, you'd expect many more of the fibers to have been recruited to share the burden before we reach exhaustion—but that doesn't seem to be happening.[6] "It's a simple disproof of the prevailing theory," Noakes told me. And while there is abundant evidence that lactic acid accumulates during exercise, it has been difficult to prove that it weakens and fatigues muscles in the ways we once assumed—with some studies suggesting that it may actually improve the muscles' movements during times of extreme exertion.[7] Given these findings, it's very difficult to pinpoint any bodily change that can convincingly explain the rapid onset of exhaustion.

That's not to mention the striking psychological effects long noted by athletes and coaches. Careful experiments have confirmed that athletes perform consistently better when they are in a head-to-head competition compared with when they are training alone. They seem to draw on some kind of hidden reserve that is only activated in certain contexts, which is hard to explain if exhaustion is only the result of depleted glycogen and accumulated lactic acid.[8]

Perhaps most damningly of all, the biochemical theory cannot explain the puzzling fact—noted by Mosso and replicated more recently—that intellectual effort alone can lead to markedly worse subsequent physical performance. In 2009 researchers at Bangor University found that cyclists experienced a 15 percent reduction in stamina after they had performed a grueling 90-minute test designed to tax their memory and concentration.[9] True, the brain consumes glucose, but it seems highly unlikely that a purely intellectual exercise could have such a large effect on physical exhaustion if feelings of fatigue were due solely to the depletion of the muscles themselves.

These enigmas have led a growing number of sports scientists like Noakes to return to a "psychobiological" theory of exhaustion that fully accepts the brain's role in determining our physical limits, just as Mosso had proposed a century ago.[10] In their view, the brain uses its previous experience, physiological sensations such as our core body temperature, its current mood and sense of mental strain, and its predictions of the remaining task to carefully judge how much exercise we are capable of performing and at what intensity. These calculations will determine how many muscle fibers to recruit and the intensity of the movements that the body can sustain, and if the brain senses that we risk overexerting ourselves, it will put the brakes on our movements, inhibiting the signals sent to our muscles and creating an overall sense of fatigue that makes it harder and harder to keep going.[11] Although that feeling of exhaustion is uncomfortable in the short term, it helps us to preserve some energy for later on and keeps us from pushing ourselves to the point of injury.

The brain's estimates of what we can achieve are generally very conservative, and that makes evolutionary sense: unless we are facing a life-or-death threat, it's generally better to play it safe to avoid potential damage. But these predictions need to be flexible to adapt to changing circumstances, meaning that it is often possible to release some of those hidden reserves with small psychological nudges. Consider a study by R. Hugh Morton at Massey University in New Zealand. In the late 2000s he asked a group of cyclists to take three identical rides during which they were asked to cycle as hard as they possibly could for a few minutes—to the point of exhaustion. In one of the trials the participants' clock was completely accurate; in the others the clock was wonky, running either 10 percent too fast or 10 percent too slow. If the brain's predictions played no role in our feelings of fatigue, the difference on the clock should have had no effect on their endurance. In reality, their stamina increased by 18 percent when the clock ran slow, and it fell by around 2 percent when it ran fast, compared to the accurate timekeeping. The wonky time perception had led the participants' brains to estimate that they had exerted more or less effort than they actually had—and adjusted their sense of exhaustion accordingly.[12]

Similar benefits could be seen by getting cyclists to compete with themselves on a virtual track that showed their current pacing alongside a previous performance. Unbeknownst to the participants, the avatar representing their previous performance had been programmed to ride faster than their personal best, and by resetting their expectations of what they could achieve, the athletes were able to exceed their previous limits.[13]

Since the prediction machine constantly updates its calculations with feedback from the body, you can also boost performance by changing the interpretation of those internal signals. Athletes usually find it harder to exercise when they are hot, for example, with the brain creating the sense of exhaustion to avoid the body overheating—but this can be manipulated. British cyclists riding in hot, humid conditions had significantly greater stamina if they were told that their core body temperature was slightly lower than it really was.[14] Similarly, a study from 2019 gave cyclists falsely high readings of their heart rate played through headphones. The feedback led their brains to overestimate their bodies' level of exertion, generating greater feelings of exhaustion more quickly.[15]

Our understanding of the psychobiological model of exhaustion is still growing, and there is an increased interest in its neural origins. By placing electrodes on the scalp as people perform exercise, researchers have started to locate the brain regions involved in processing our expectations of exercise and creating the sense of exhaustion. At the center of it all is the prefrontal cortex—lying behind the forehead—which uses our factual knowledge of the exercise at hand, our previous experiences, and sensory signals from across the body to predict its remaining physiological budget and the consequences of its exertions. It then transmits these calculations to the motor cortex (which plans our movements) to control our energy output accordingly—and to stop us from continuing our exercise when we are at risk of causing bodily damage.[16]

If he were alive to see this research, Mosso might have considered these regions to be the seat of "the will." But they are all components of the same prediction machine that controls so much of our physical reality.

This new theory of exhaustion, one that rightly places the brain as controller of what the body can do, helps us to understand the influence of placebo treatments in sport. If we consider Virenque's amazing time trial in the 1997 Tour de France, the injection of the "magic potion" increased his perception of what he could achieve. His brain calculated that it could devote more of the body's resources to the race without risking injury, allowing his muscles to work harder on the track. The fact that it was only sugar water didn't matter: because of its effects on the prediction machine, the injection still increased the amount of energy that Virenque was able to expend. We may describe the substance as "inert," but in terms of its effects on performance, it was anything but. Virenque's belief and the sense of ritual accompanying the injection imbued the substance with power.

Carefully controlled studies show that a large proportion of all commercial sports supplements may aid performance by enhancing someone's perceptions of their abilities, independently of any direct physiological effects.[17] Caffeine, for instance, has long been considered a muscle stimulant that can enhance performance in many sports—yet that is largely a product of our expectations about what it can do. In one study, student bodybuilders were given a shot of a bitter-tasting liquid, which they were led to believe contained a high concentration of caffeine. In reality it was a dose of decaf—but they still managed to increase the number of repetitions by around 10 percent above their previous limit.[18] People who have taken caffeine under the assumption that it is an inert substance, meanwhile, tend to see a much smaller performance boost.[19]

Expectation effects may even lie behind some of the benefits of banned drugs, including anabolic steroids and erythropoietin, a hormone that stimulates the production of red blood cells. In a 3-kilometer race, runners taking an inert saline injection raced 1.5 percent faster than a previous personal best if they believed they had taken a substance similar to erythropoietin—a slim but significant advantage that could easily give them the edge in a close race, given that Olympic rankings can often depend on a difference of a fraction of a second in race times. In other words, athletes such as Virenque may have no need for the doping that risks their careers if they are able to change their expectations through other means.[20]

Coaches administering a placebo to an athlete under the perception that it is an illegal drug may be a morally dubious practice. But some scientists are concerned that coaches may find even smarter ways to bend the rules around doping. It is possible, for instance, to enhance placebo effects by using a banned substance during training, then slowly changing the dose until the drug has been completely replaced by an inert substance. The athlete will enter the competition with inflated expectations of their success, and have a vast physical advan-

aerobic capacity (also known as $VO_2$ max) is the peak rate of oxygen consumption over a thirty-second period during this time—and it is meant to reflect how well the lungs and the heart are able to deliver fuel to your muscles. The higher your $VO_2$ max, the better your endurance during exercise.

To find out if positive feedback could alter this basic measure of fitness, Jeff Montes and Gabriele Wulf at the University of Nevada, Las Vegas, asked a group of participants to undergo two $VO_2$ max tests. Although the first test was accurately measured, some of the participants were given false positive feedback. In a casual conversation, they were told that their score was better than most other members in the group, while those in the control group were not given this encouraging information. Within a few days, they took the $VO_2$ max test again, and those with the enhanced expectations scored significantly better, while those in the control group performed slightly worse. Overall there was a roughly 7 percent difference between the groups. In other words, how fit someone appeared—according to the standard test of aerobic capacity—changed according to how fit they thought they were.[21]

Besides increasing aerobic capacity, heightened expectations of this kind can also improve the efficiency of a runner's movements. In another experiment, the participants were asked to run on a treadmill at a fixed rather than increasing speed for ten minutes. The researchers found that raising the participants' expectations of their abilities led to a significant decrease in oxygen consumption during the exercise. This suggested that the muscles were burning less energy to maintain the same pace.[22] That's an important change that should, in turn, have left them with greater resources later on, should they have needed them—enhancing overall endurance. Perhaps because of their reduced fatigue, these participants were also more likely to experience a mood boost after exercise.

Astonishingly, our expectations of our physical abilities may override certain genetic dispositions for exercise, according to a paper published in 2019 in one of *Nature*'s prestigious journals. The scientists first performed a genetic test to identify whether their participants were carrying a certain version of the CREB1 gene, which, previous studies suggest, can reduce people's aerobic capacity and increase their body temperature during exercise—making the whole experience harder and more unpleasant. The test was real, and the researchers kept a record of the results. The outcome given to participants, however, was completely random, creating expectations that they either were or were not "naturally" good at exercise. And that information had an important effect on their physical endurance, with negative expectations reducing air flow in and out of the lungs and the transfer of oxygen and carbon dioxide—resulting in lower overall stamina. Importantly, the effects of the expectations appeared to exert more of an influence than did the actual gene type on some of these physiological measures. When it came to the exchange of oxygen and carbon dioxide, for instance, a participant's belief that they were genetically indisposed to exercise proved to be more damaging to their performance than the presence of the actual gene variant.[23]

We can't all rely on scientists giving us false feedback, of course, but there is some evidence that we can bring about similar changes ourselves, without any deception. Cyclists improved their performance after using a sports supplement, for instance, even if they had been told that it was physiologically inert before they took it.[24] In this case, the supplements seem to work like the open-label placebos that have proven to be so effective for pain relief. An understanding of the brain's potential to control physical performance was enough to bring about a boost. So feel free to use any aids that work well for you. Whether it's a favorite drink, fancy sportswear, or motivational music, the change in mindset will bring the benefits.

Grace Giles at the US Army's Combat Capabilities Development Command Soldier Center, meanwhile, has shown that reappraisal techniques can reduce people's perceptions of exertion as they go about their exercise, so that it feels less exhausting.[25] As we've already seen in Chapter 3, reappraisal involves a slightly more dispassionate examination of our feelings and an effort to consider whether they might be neutralized or even interpreted in a more positive light.

Many of us start forming negative thoughts about exercise before we even walk out the front door, so an important first step is to focus on the immediate benefits you want to get from this exercise—such as feeling refreshed and energized at the end of the workout. Further into the exercise it can be easy to interpret feelings of exertion—such as breathlessness or aching muscles—as a sign of poor fitness. You may start to assume that this is proof that you just are not cut out for exercise, and the more you focus on this thought, the worse you feel. In this case, try to consider whether these sensations are, in fact, desirable. In just the same way that we can reinterpret the side effects of drugs as a sign that a medication is taking effect, we can rethink our aches and pains as evidence that our exercise is really changing our body. If you are breathless and your limbs are starting to feel heavy, that's a sign that you are strengthening your muscles, expanding your lungs, and increasing your heart's stamina. The exercise is working.

Once you start exercising more regularly, you might still face days of frustration—when you just don't seem capable of running as fast as you want or lifting as much weight as you would like. Rather than dwelling on the feelings of failure, however, you can remind yourself that doing any exercise at all is better than nothing; your body is still benefiting. Perhaps you need to recover from a hard week's work, or from other stress in your life. That realization will make the session feel a lot less tiring than if you keep on agonizing over your poor performance and beating yourself up for not reaching your goal.

You should still be mindful of potential strain or overexertion, of course, so be sure to test your abilities in small steps and to consult your doctor if you have any concerns about your safety. The aim is to avoid overinterpreting your struggles as a reflection of an inherent inability, focusing instead on the overall arc of progress through incremental steps. Research shows that recognizing that your physical fitness is within your control, and can be improved over time, will ensure that you maintain your enthusiasm and energy, rather than descending into self-defeating rumination—a seemingly obvious fact that many people nevertheless forget.[26]

As a formerly reluctant exerciser myself, I have found that this kind of reframing really does help to take the pain out of a workout. I hated PE as a child, but knowing the importance of physical activity, I have tried to work out regularly for years. Yet it always felt like a burden; I often couldn't wait to get off the treadmill. Learning to reframe the feelings of exertion, however, has helped me to feel much more energized during and after my workouts. When I feel like I am about to hit the wall, I find it particularly helpful to remind myself that my body has hidden reserves of energy that can be tapped and to imagine my lungs expanding and my heart pumping more nutrients to my limbs. And during the workout I make an effort to regularly remind myself of the long-term benefits that exercise can bring. In addition to regular cardio workouts, I now do high-intensity interval training five times a week—and it's genuinely the highlight of my day. I can only describe the change of mindset as a big release, allowing my body to perform the exercise that it was always capable of doing.

## INVISIBLE EXERCISE

With these techniques we can all begin to ease ourselves into a more active lifestyle. The power of reframing does not end in the gym, though. Many everyday tasks can strengthen the body, even though they look nothing like a typical workout, and according to some groundbreaking research, the meanings that we attach to those activities may determine whether or not we reap the full benefits of the exercise.

The existence of "invisible exercise" should not be a surprise—our understanding of it dates back to the very first study to examine the benefits of physical activity. Soon after World War II, Jeremy Morris at the UK's Medical Research Council wanted to understand why some people are more prone to heart disease than others. Suspecting that exercise could be the answer, he looked for a group of people of a similar social class and status whose professions differed only in the amount of time they spent being physically active.

Men working on London's double-decker buses proved to be the perfect population to study. Although their educational and financial backgrounds were roughly the same, the drivers spent most of their day sitting, while the conductors were constantly active, climbing up and down the stairs to collect fares, issue tickets, and help passengers with their luggage. In total, the average conductor climbed around 500–750 steps every day.[27] Although this was relatively gentle exercise—compared to training for a marathon, say—Morris found that the daily activity roughly halved the bus conductors' risk of heart failure.

Morris became known as "the man who invented exercise," and his findings inspired an avalanche of further research on the benefits of exercise. The much-touted recommendation that we should aim for 150 minutes of *moderate* exercise (or 75 minutes of *vigorous* activity) per week can be traced back to those London bus conductors. These guidelines are regularly publicized—but many of us are much less clear about what actually counts as moderate or vigorous exercise, and that is important when it comes to the formation of our fitness mindsets. To compare the intensity of different activities, physiologists use a quantity known as "metabolic equivalents," or METs—which is the metabolic rate of the exercise divided by the metabolic rate of resting. If an activity is 2 METs, for instance, you are burning twice as many calories as if you were sitting and watching TV. Moderate exercises are between 3 and 6 METs, and vigorous exercises, anything above 6 METs. It doesn't matter whether you get this exercise in short bouts or in a single session—it's the total time over the week that matters. And many everyday activities and pastimes meet these requirements. Just consider the following table:[28]

| ACTIVITY | METABOLIC EQUIVALENT |
|---|---|
| **Housekeeping** | |
| Vacuum cleaning/Washing the floor | 3 |
| Cleaning windows | 3.2 |
| Making the bed | 3.3 |
| Cooking/washing up | 3.3 |
| Moving furniture | 5.8 |
| **DIY** | |
| Carpentry (e.g., hammering nails) | 3 |
| Painting/wallpapering | 3.3 |
| Roofing | 6 |
| **Gardening** | |
| Trimming shrubs | 3.5 |
| Chopping wood | 4.5 |
| Mowing the lawn | 6 |
| **Pleasure** | |
| Walking the dog | 3 |
| Drumming | 3.8 |
| Outdoor play with children | 5.8 |
| Dancing | 7.8 |

How many of us mow the lawn, play with our children, or dance the night away at a club without even realizing that we're really working out? Even the daily commute could count. A study from Imperial

College London has shown that roughly a third of all English people using public transport already meet the governmental guidelines for physical activity from their commutes to and from work—by waiting for buses, walking to or from the station, or changing trains.[29]

At the very least, a greater appreciation of these kinds of activities should lead us to be more positive about our level of fitness—a changed expectation that could reconfigure the prediction machine so that other, more formal workouts feel like less of a strain. Even more remarkably, however, this shift in mindset might also determine the long-term benefits of the activities themselves, according to a study by Alia Crum and Ellen Langer at Harvard University. By thinking of everyday activities as exercise, rather than work, it seems, we can become healthier.

As you might recall from the introduction, the participants in this study were cleaners from seven different hotels. Crum and Langer suspected that few of these cleaners would be aware of the sheer amount of exercise that their job entailed, and, given the power of expectation to shape our physiology, that this might prevent them from gaining the full benefits of their daily workout. To test the idea, the scientists visited four of the hotels and gave the cleaners information about the kinds of physical activity that count as exercise, emphasizing that it "does not need to be hard or painful to be good for one's health . . . it is simply a matter of moving one's muscles and burning calories." They then offered some details about the energetic demands of the cleaners' own work—that changing linen for 15 minutes burns 40 calories, vacuum cleaning for 15 minutes burns 50 calories, and cleaning bathrooms for 15 minutes burns 60 calories—which, over the course of the week, should easily add up to the US surgeon general's exercise recommendations. Besides offering flyers containing these facts, the researchers also placed posters conveying the information on bulletin boards in the cleaners' lounges, so that they would have a daily reminder of the exercise they were getting.

A month later, the scientists visited the cleaners again to measure any changes in their health. Despite reporting no alterations to their diet or increased physical activity outside work, the cleaners who received this information had lost about two pounds each, and their average blood pressure had dropped from elevated to normal. The shift in expectation—and the meaning that the cleaners ascribed to their work activities—had changed their bodies, while the cleaners at the remaining three hotels, who had not received the information, showed no difference.[30]

It was, admittedly, a relatively small study—and there was always the possibility that after they'd been given the information, the cleaners had put a bit more "oomph" into their work. But a follow-up by Crum, who is now at Stanford University, and her colleague Octavia Zahrt provides much more compelling evidence that people's expectations really can influence the long-term benefits of exercise through the mind-body connection. Their study used data from health surveys monitoring more than sixty thousand people for up to twenty-one years. Crum and Zahrt found that the "perceived physical activity" of the participants—whether they felt they did more or less exercise than the average person—could predict their risk of mortality, even after the researchers controlled for the amount of time the subjects said they actually spent exercising, and other lifestyle factors, like diet.

Importantly, some of the participants in these surveys had worn accelerometers for part of the study period—yet the influence of their perceived physical activity remained after the researchers had taken these objective measures of physical activity into account. Overall, people who took a more pessimistic view of their fitness were up to 71 percent more likely to die during the surveys, compared with those who thought they were more active than average—whatever the status of their actual exercise routine.[31]

As a science writer I was initially amazed when I first heard of this paper—but the more I've dug into the science of expectation, the less surprising these results seem. We have seen, after all, how things like blood pressure can change due to our expectations of a pill's effect. If our expectations of a beta-blocker can have a noticeable effect on our health, why should our perceptions of our physical fitness—which we carry with us in every activity, every day—be any less important? When we put it like that, the really surprising thing is that it has taken researchers so long to investigate the possibility.

We now know that many of the other benefits of exercise can be the product of expectation. Exercise is known to improve people's mood and mental health, for example, and it also acts as an analgesic, reducing the symptoms of both acute and chronic pain. Both the mood boost and the analgesia are thought to arise from the release of endorphins. While that may be an automatic physiological reaction to physical activity, people's beliefs appear to play a large role in triggering the response—and educating people about that potential seems to enhance the effects.[32] If you expect to feel more relaxed and energized, or for your aches and pains to evaporate, you're more likely to experience that relief.

Could there be a danger that we could take this message too far? If we begin to focus too much on reappraising our existing activities and improving our opinions of our existing fitness, couldn't people become complacent—and make even *less* effort to get the exercise they need? Fortunately, the studies so far suggest that this isn't likely to happen. You can encourage people to take a more positive view of their fitness without pushing them into indolence.[33] Governments would do well to remember that when they design health campaigns encouraging exercise. According to this research, strict or judgmental language—which emphasizes the population's current lack of fitness—will backfire, compared to messages that encourage people to take a more optimistic approach. Instead, scientists like Crum and Zahrt argue the messaging should reiterate the fact that even small improvements can have significant long-term effects. The recommendation of thirty minutes of moderate exercise a day, five days a week might be the gold standard, but even fifteen minutes a day can increase your life expectancy by three years.[34]

More generally, Zahrt and Crum's studies suggest that we should avoid "upward comparisons," constantly judging ourselves against the people who are fitter than we are. While there's nothing wrong with a bit of aspirational thinking, it can easily turn into feelings of inadequacy, and we can come to form more negative expectations of our fitness. Such perceptions may then reduce the benefits of our workouts.

This is especially important to remember when we consider our social media feeds. Instagram and TikTok are full of "fitspiration" or #fitspo accounts, with Photoshopped images of toned bodies working out. The videos and photos are meant to be motivational, but a study published in 2020 suggests they do more harm than good. The participants (all female undergraduates in Adelaide, Australia) were first asked to scroll through a set of images—either attractive shots of exotic travel locations or eighteen pictures of fitness gurus completing their workouts. The participants then spent ten minutes on a treadmill, exercising at the speed of their choice, and completed a series of questionnaires about their feelings. Those viewing the fitspiration images suffered on almost every measure. They had worse body image and experienced greater feelings of fatigue during exercise; and rather than experiencing the "runner's high," they were in a significantly worse mood following the workout compared to those who had seen the travel pictures.[35]

The pictures appeared to have damaged the participants' perceptions of their own fitness, and the negative self-comparisons led them to believe that they were less healthy than they really were. And the resulting sense of inadequacy rendered the exercise more arduous and less enjoyable—completely negating any of the claimed motivational benefits.

The need to form positive but realistic personal goals is especially important when we consider one more remarkable way to prime our mind and body for greater fitness and health. Using nothing more than our imagination, we can tweak the brain's predictions to increase the strength of our muscles and finesse our physical performance.

## WORK THE MIND, WORK THE BODY

With twenty-eight medals to his name (twenty-three of which are gold) the US swimmer Michael Phelps remains the most decorated Olympian of all. Phelps's abilities appear to defy the limits of the human body, and some journalists had questioned whether his feats were just "too good to be true." Yet Phelps voluntarily signed up for many anti-doping tests during his career and passed them all.

Perhaps his amazing racing times are better explained by another unnatural advantage—his extraordinary powers of visualization. During training, and in preparation before a big event, he would imagine the perfect race. "I can see the start, the strokes, the walls, the turns, the finish, the strategy, all of it," he wrote in his autobiography, *No Limits*.[36] "Visualizing like this is like programming a race in my head, and that programming sometimes seems to make it happen just as I imagined it." It is this capacity—rather than a purely physical ability—that helped him to become the greatest Olympian, he believes.

Scientific experiments have confirmed that the effects of visualization can be profound for professional sportspeople and casual exercisers alike.[37] The most striking—and surprising—effects have been found in people's muscular strength. In one study scientists measured participants' forearm strength before they conducted a form of mental training. The task was easy, if boring: they had to spend fifteen minutes a day, five days a week, imagining that they were lifting a heavy object, such as a table, using their forearms. Some were told to do this from an internal perspective, picturing the movements as if they were lifting the weight themselves; others were asked to do this from an external perspective—as if they were viewing themselves from outside their own body. The control group performed no practice at all.

Six weeks later the results were staggering, with the first-person internal visualization producing an 11 percent increase in strength—despite this group not having physically lifted a single weight.[38] Those using the external perspective saw a more modest improvement of around 5 percent (though the researchers could not be sure that this was statistically significant), while the control group actually seemed slightly weaker.

Like the effects of the placebo supplements, these findings would be inexplicable if strength were determined solely by muscle mass.[39] With the new psychobiological view of exercise, however, it makes perfect sense. Remember that performance relies on the brain's expectation of what the body can achieve and how arduous an exercise will be, which the brain then uses to plan the force and exertion of muscles. Mental imagery allows you to consciously refine these predictions and increase the body's perceptions of its own abilities, boosting the signals that it will send to the muscles and improving the coordination of movement. As Noakes's work has shown, athletes do not normally

Change in arm strength after six weeks' mental training

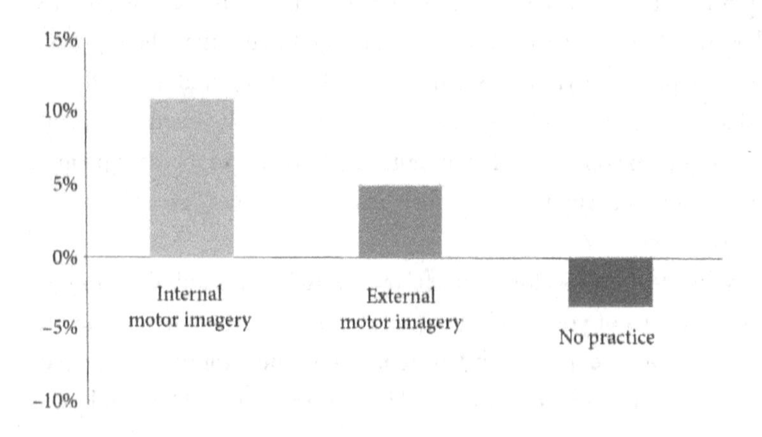

recruit the majority of their muscle fibers even during peak exertion, but the imagery might be encouraging the body to summon up more of the fibers that have been left unused.

Brain scans of athletes engaging in vivid visualizations of their events show that visualizing exercise activates areas of the primary motor cortex and basal ganglia that are typically involved in planning and executing movements, as the brain calculates exactly which muscles need to be stimulated and the effects that exertion will have on the body.[40] And these upgraded expectations will then translate to a real performance boost. According to this theory, internal imagery is more successful than external imagery because it leads you to make more detailed predictions of the way you will feel during the exercise, so that the body can execute the movements more effectively.

Mental practice cannot, and should not, replace physical practice whenever possible, of course, but it allows athletes to make the most of their rest periods, and to avoid a loss of strength after injury.[41] People's muscles normally weaken when their limbs have been placed in a plaster cast, for instance, but scientists from Ohio University have found that a few minutes of mental practice a day can reduce those losses by half.[42] For the rest of us, mental practice should just be another tool to maximize the benefits of our workouts. If you find that going to the gym is a strain and you want to change your mindset about exercise, then regularly visualizing the benefits could make the whole process more appealing. Various studies—in adolescent, middle-aged, and elderly participants—have now shown that regularly practicing mental imagery of exercise for a few minutes each week can increase people's motivation and enjoyment of their exercise regimes, as well as their performance.[43]

When testing this for yourself, try not to be overambitious in your visualizations of your performance. You don't want to set yourself up for disappointment—which will reduce your motivation—or prepare yourself for overexertion that could result in injury. (The mind-body connection can achieve only so much without constant physical training.) As you visualize your exercises, try also to focus on the positive sensations that you hope to feel during exercise; the feeling of being energized and buzzing rather than fatigued or exhausted, for example. Just like Phelps, you will be "reprogramming" the mind-body connection, allowing yourself to overcome the mental restraints that might have been hampering your performance, so that exercise no longer feels like an insurmountable challenge.

## HIDDEN STRENGTH

We still don't know the full extent of the brain's influence over physical performance—but anecdotal evidence suggests that it may be truly immense. In 2012, for example, Alec Kornacki was crushed by his BMW 525i while changing a tire—only for his twenty-two-year-old daughter, Lauren, to lift the car to release him. "It was like a table with a short leg. It kind of balanced it back out and shifted enough to free my dad," she told ABC News.[44] She went on to perform CPR—and saved his life.

The phenomenon of ordinary people showing incredible abilities in times of crisis is known as "hysterical strength," and it has been reported in people of all ages, from two teenage girls who released their father from a three-thousand-pound tractor, to a septuagenarian who raised a Jeep to save his son-in-law.[45] If these events sound like the actions of the Hulk, well, it's not exactly a coincidence. The creator of the comic strip, Jack Kirby, was apparently inspired to create the character after seeing a mother lift a car off a child, when sheer panic released hidden reserves of strength.[46]

Lifting a car, even a relatively small distance, would normally lie beyond the ability of the keenest bodybuilder. So what was going on? These astounding feats are usually explained by an adrenaline rush, but according to some scientists, they can be better explained by an explosion of energy emerging from the brain. While the prediction machine usually compares its resources with the demands of the situation and carefully calculates how much exertion the body can afford without risking total depletion or injury, sheer emotional urgency can override the brain's typically cautious control; it essentially decides that the task is so important, it is worth the risk of injury. As a result, the brain begins to fire up more of the limbs' muscle fibers, producing an incredible explosive force.[47]

Displays of hysterical strength are dangerous: torn muscles and broken teeth are the usual consequences of these events. And that's precisely why the brain is normally so careful in parceling out our efforts and limiting our exertion, even when athletes are in a serious career-defining competition. But these anecdotes nevertheless provide the startling reminder that our physical abilities are often limited by the brain as much as by the body. We don't often need to be able to lift a car, but most of us could do with a bit of a helping hand in our fitness regime. If we can use expectation effects to plumb even a small portion of the physical reserves shown by Virenque, Phelps, or Kornacki, we can all enjoy a fitter future.

### HOW TO THINK ABOUT . . . FITNESS

- Before you exercise, think carefully about your aims for the session. How do you hope to feel at the end? And what are your performance goals? You might hope to set a new personal best, or you might be looking for a quick mood boost—either way, you'll increase your motivation and help to calibrate the prediction machine for the oncoming activity if you define what you want from the activity before you start.

- Embrace any of the mental crutches that help you to feel good about exercise. Certain foods, drinks, clothes, or songs will lead you to feel energized. As we saw with the open-label placebos, you can know that the advantages come from belief and still benefit—so use aids with the best personal associations and allow your expectations to bring the performance boost.
- Question your assumptions about your innate disposition toward sports and exercise. Remember that your expectations can be more important than known genetic factors in determining your physiological response to a workout.
- Reframe feelings of exertion and effort. Moderate aches and pains and sensations of fatigue are proof that you are strengthening your body. Remembering this can make the whole workout feel more enjoyable and less exhausting.
- Recognize the physical activity you perform outside your regular workouts—during housework, commuting, or working on your hobbies. (You may even want to keep a diary for a week.) Thanks to an expectation effect, you can optimize the physiological benefits of this exercise—simply by paying more attention to it.
- Avoid "upward comparisons"—judging yourself harshly against other people—since this will lead you to form more negative impressions of your level of fitness.
- During rest periods, spend a few minutes visualizing yourself performing the exercises of your next workout. This will increase your muscle strength and prime your brain for better performance.

# 6

# THE FOOD PARADOX

## Why indulgence is essential for healthy eating

Imagine you are considering a new diet to cut your calorie intake and to curb your tendency to snack between meals. Of the following two daily plans, which is the one more likely to leave you feeling full and satisfied? And which is more likely to help you lose weight quickly?

### The Super-Slimmer

*Wholesome meals for a healthy future*

Breakfast

Two whole-grain slices of smashed avocado toast
Mango and pineapple smoothie (no added sugar)

Lunch

Pole-and-line-caught tuna niçoise salad
Fresh glass of orange juice

Dinner

Braised low-fat organic chicken and asparagus

Optional post-workout revival snack

No-thrills healthy granola bar

Or

## The Bon Viveur

*Delicious and decadent meals to maximize pleasure*

BREAKFAST

Full-butter croissant

Mexican chile hot chocolate

LUNCH

Spaghetti with spicy puttanesca sauce (tomatoes, anchovies, and olives)

Mixed fruit salad (pineapple, orange, melon, mango, apple, and blueberries)

DINNER

Fish pie with creamy mashed potato crust

Mixed leaf salad

OPTIONAL POST-WORKOUT REWARD:

Two donut bites

If you have ever been on a diet before, the Super-Slimmer probably seems like the best choice for quick weight loss. Without the post-workout reviver (and assuming standard portion sizes) it would amount to around 1,750 calories per day—a decent reduction for the average person, and enough to lead to steady weight loss.[1] The downside, of course, is that it might come with less overall satisfaction.

The Bon Viveur, on the other hand, seems packed full of calories. It starts out with a croissant and hot chocolate, includes spaghetti for lunch, and finishes with a pie! Surely that menu can't supply less energy than the niçoise salad and the chicken braise? You might choose this option if you want to enjoy life—but you might not expect it to help you lose weight quickly. Yet it has even fewer calories than the Super-Slimmer—just 1,632 in total if you don't include the post-workout snack.[2]

When you *do* include the post-workout treats, the mismatch between our expectations and reality is even starker. The granola bar, which sounds like a "sensible" snack, is so heavily sweetened that it contains 279 calories overall—more than twice the 110 calories of the two minidonuts.

If you are surprised by these figures, you are not alone: surveys show that most people find it difficult to gauge the number of calories in foods, and we are particularly prone to underestimating the nutritional content of products that are typically marketed as health foods with slogans like "simple," "wholesome," and "guilt-free."

The most obvious consequence is that we may *consciously* decide that we have a license to snack more if we think we've consumed less than we really have. The true effects may go much deeper, however. Because of the influence of the brain's predictive processing, our expectations of a food's nutrients will also directly influence our body's responses to the food, including digestion (the breakdown and absorption of nutrients in the gut) and metabolism (the use of that fuel to power our cells). When we think we are eating fewer calories than we actually are, the body responds as if that's the truth: it feels less sated so that we experience much worse hunger pangs, and it stops burning so much energy to preserve its existing fat stores. We are experiencing a "deprivation mindset," which may make it much harder for us to lose weight on a seemingly spartan diet than when we are eating meals packed with our favorite comfort foods.

Whatever particular diet plan we are following, this expectation effect has the potential to make our weight loss far harder than it needs to be. If we want to maintain a healthy weight, we don't just have to change what we eat; we have to change our whole way of thinking and talking about the food we're ingesting. And a key part of that is to avoid seeing "healthy" and "pleasurable" as a dichotomy and to recognize that a sense of indulgence should be an essential ingredient in every meal.

## IN SEARCH OF LOST MEALS

To understand the ways that our brain's predictions can influence hunger, digestion, and metabolism, we must first examine the famously ravenous appetite of one of neurology's most famous patients, Henry Molaison.[3] Born in Connecticut in 1926, Molaison was a healthy middle-class boy until in late childhood and early adolescence his parents and teachers noticed that he would often "zone out" midconversation for about ninety seconds, with an absent look on his face. Doctors diagnosed a form of epilepsy, and beginning around Molaison's fifteenth birthday, the attacks became far more violent, with rhythmic convulsions across his body that left him shaking and writhing on the floor.

Epileptic seizures are caused by sudden bursts of electrical activity that stop brain cells from communicating with each other. When Molaison failed to respond to medication, his team decided to operate, removing a portion of each of his temporal lobes, where the attacks were thought to originate. It worked: Molaison no longer experienced the severe seizures that had plagued his life, but it soon became clear that this relief had been achieved at a huge sacrifice. Although Molaison could remember events from his past, he had completely lost his ability to form new recollections. In the hospital, for example, Molaison would meet the same staff time and again without remembering ever having seen them before. You could tell him a surprising fact in the morning, and he would be equally dumbfounded by the discovery in the afternoon, as if he were hearing it for the very first time; he was, in the words of the neuroscientist Suzanne Corkin, living in a "permanent present tense."

Over the subsequent decades, studies of Molaison—known in the medical literature by his initials, H.M.—completely revolutionized our understanding of the way the brain works. He enabled scientists to link memory formation to an area of the brain known as the hippocampus, which had been severely damaged in his operation, and showed that we can learn some skills nonconsciously even when we have no explicit memory of the learning event. Few people have had such an influence on neurology and psychology as Molaison, who died in 2008, and he is now famous to science students across the world.

What is much less well known, however, is his contribution to the understanding of appetite. The scientists studying Molaison had long noted that he rarely reported being hungry, and yet he always seemed ready to eat.[4] In the early 1980s Nancy Hebben at Harvard University and colleagues decided to put his appetite to the test by asking him to rate his satiation on a scale of 0 (famished) to 100 (completely full) both before and after his meals. If our appetite were mostly directed by signals from the stomach, you would expect the rating to rise after the meal; Molaison's memory deficit should have had no effect on how full he felt. Yet Molaison gave the same ratings—around 50—at both points. Because he was stuck in the "permanent present," his hunger never seemed to change.

To see whether his memory deficit would also change his eating behavior, the scientists performed a dinnertime experiment. After Molaison had eaten a meal, the staff at his care home cleared the table, and within one minute they gave him a second meal. Amazingly, he ate almost all of it, leaving only the salad. Even then, he showed only a moderate increase in satiety, whereas most people would experience a complete loss of appetite after two substantial meals.[5] Without a memory of what he had eaten, he seemed to have no way of regulating his food consumption.

Molaison might have been a one-off, of course, but studies of several other amnesiacs have since come to similar conclusions. "It's really amazing to see," said Suzanne Higgs, who has conducted some of this research at the University of Birmingham in the UK. She remembers one patient looking at the clock when she asked him if he wanted to eat. "It's like he couldn't really figure out whether he was hungry or not, and that was his only way of knowing whether it was appropriate to eat." Another amnesiac participant (not Higgs's) proved to be so ravenously hungry after two large meals that he was ready to eat a third plateful; fearful of the health impact of letting him eat so much food, the scientists decided to remove the plate after he had consumed a few mouthfuls.

How could this be? There is no doubt that appetite comes partly from the activity in the digestive system, the "bottom-up" sources of information. When we eat, the gut starts to stretch to make room for the food. We have sensors in the muscles surrounding the esophagus and gastrointestinal tract that can detect this movement. They pass their signals through the vagal nerve into the brain and help to create a feeling of reward and satisfaction when we are full (or a bloated feeling when we have eaten too much).[6] The gut also has its own chemical receptors that can detect the presence of nutrients, such as fat or protein, and when they are stimulated they release hormones that curb our hunger.[7]

The experiences of amnesiac patients like Molaison, however, suggest that these sensory cues can give us only a crude estimation of how much we've eaten. It seems that the prediction machine also has to draw on other "top-down" sources of information—such as memory and expectation—to make sense of information coming in from the gut and to create the sense of hunger or satiation. Without the capacity to form a memory of the day's meals, Molaison's brain wasn't able to contextualize bodily signals in this way, meaning that he would never feel fully sated after a meal.

You might reasonably wonder about the relevance of these findings to your life. But you don't need to have brain damage to suffer a poor memory, and even mild forgetfulness appears to cause overeating.[8] What's more, researchers like Higgs have shown that even small changes to the ways we think about food—past and present—can change the brain's assessment of what it has eaten, with profound effects on our appetite.

In one notable experiment, Higgs invited a group of students into her lab after lunch to perform a taste test on some cookies—which, after completing a couple of questionnaires, they were free to consume. Higgs found that prompting the subjects to remember their lunch—by spending a few minutes noting down their recollection of what they had eaten—reduced their total consumption of cookies by about 45 percent, compared to participants who wrote about their general thoughts and feelings, rather than their food memories from that day. It was a difference of around four cookies per person. This was not the case for students who wrote about a meal from the day before—a more distant event that would have had little effect on their feelings of satiety at that moment. Instead, it was expectations of current satiety, based on recent memory, that seemed to matter.[9]

The role of memory and expectation in creating a sense of satiety also explains why the appearance of food can inordinately influence how much we eat. In 2012 a team at the University of Bristol first showed their participants a bowl of either 300 milliliters or 500 milliliters of cream of tomato soup, which they were asked to eat. Unbeknownst to the participants, however, the bowl was rigged with a small pump that could either increase or reduce the amount of soup they consumed. As a result, some participants believed they were eating 500 milliliters of soup—a relatively large portion—when they were actually eating a more standard 300 milliliters, and vice versa. Sure enough, the participants' hunger over the next three hours was largely determined by what they had remembered seeing, rather than the amount they had actually consumed. If they had eaten 300 milliliters but seen 500 milliliters in the bowl, they felt much less hungry than participants who had eaten more but seen less. Their sense of fullness and satisfaction was almost completely the result of their "expected satiety": that is, it was based on their visual memory of what they believed they had eaten, rather than the food they had actually consumed.[10]

Exactly the same response could be seen with students who were brought into a lab to eat an omelet for breakfast. Before completing questionnaires, they were shown the ingredients of the omelet and asked to confirm that they were not allergic to any of them. The twist was that some were shown just two eggs and 30 grams of cheese, while others were shown four eggs and 60 grams of cheese. In reality, all the participants ate a three-egg omelet with 45 grams of cheese—but the initial presentation changed their satiety and hunger for hours afterward. Thanks to a lower expected satiety, those shown the two eggs and a smaller lump of cheese subsequently ate more pasta from a buffet at lunchtime, compared with those who had seen the more generous spread of ingredients.[11]

Many of us form inaccurate food memories like these every day—with serious effects on our waistline. The unhealthy habit of working, watching TV, or surfing the internet while eating can act as a distraction that impairs memory formation of the foods we have consumed, reducing our expectation of feeling sated. "It's actually analogous to what we see with amnesiac patients, because you're not able to encode those new meal memories," says Higgs, who has investigated this phenomenon. As a consequence, we not only eat more during the meal itself, but we will eat more snacks over the next few hours, too.[12]

Then there is the presentation of manufactured foods, which can often disrupt our ability to accurately assess the contents of what we are consuming. In the past our ancestors may have had a much better idea of the ingredients that went into a dish. When we buy ready-made food and drink, however, we have little idea of the true quantities of ingredients that have gone into them. A smoothie, for example, contains many portions of fruit, but it looks much smaller in the bottle. As the brain calculates its daily intake, it remembers having eaten a lot less than it would have if it had seen the whole fruit bowl that went into the smoothie—creating the expectation of hunger later in the day.[13]

The marketing around supposed health foods can also skew the brain's estimate of what it has eaten. A food can be labeled "low-fat" even if it has a high sugar content, simply because it has slightly reduced fat compared to the standard product, for example. The result is greater hunger later on. Various studies confirm that exactly the same food—such as a serving of cookies—will lead to lower satiety when it is labeled "healthy" or "low calorie" compared with when it does not carry the label, thanks to the expectation that it will be less satisfying.[14] Indeed, the deeply embedded association between the idea of healthy eating and feelings of hunger can be so strong that eating a virtuous snack may be worse than not having anything at all. Participants who were given a sample of a "healthy" chocolate-flavored protein bar, for example, were not only less satisfied than people who had eaten the bar when it was labeled "tasty" (see below); they actually felt hungrier than people who had eaten nothing at all.[15]

Such expectation effects would be damaging enough for any dieter. But as we shall now see, the consequences do not end with our sub-

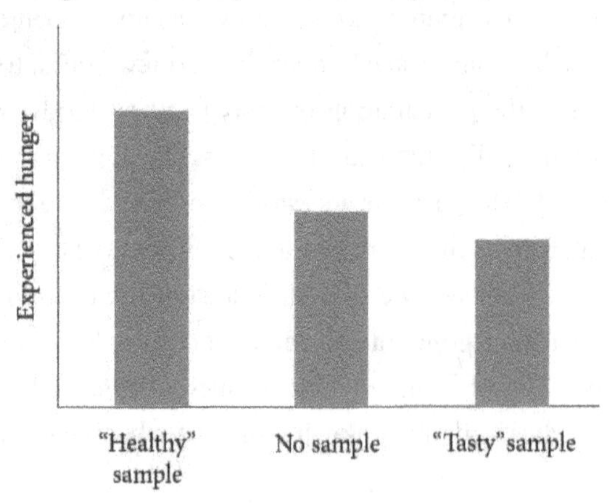

jective feelings of satiation; our beliefs about what we have eaten can also influence our digestion and metabolism. Through the power of the mind-body connection, our attitudes to food can even determine how well we absorb crucial nutrients, such as iron, that are essential for our health.

## MIND OVER MILKSHAKE

When it comes to experiments examining digestion, milkshakes are the staple. One reason is their palatability: you'll be hard-pressed to find someone who doesn't like a milkshake, particularly among the student population who make up the majority of experimental subjects. Another is the capacity to disguise the snack's contents: once ingredients have been blitzed in a food blender, it is naturally much harder to guess what went into it. This makes it far easier for scientists to manipulate participants' expectations without other factors—such as familiarity with a particular foodstuff—skewing their responses.

A notable study examined the effects of expectation on the subjects' ghrelin responses. Ghrelin is a hormone secreted by the stomach when it is empty that binds to receptors in the hypothalamus—a region of the brain that is involved in regulating many bodily functions. Ghrelin is often called the "hunger hormone" because it stimulates our appetite—its levels are highest just before we are about to eat and lowest just after a meal. But it's perhaps better seen as an energy regulator. When ghrelin levels are high, the body reduces its resting metabolic rate—so it is burning less energy overall—and starts to preserve its body fat in case of further scarcity. It can also make us lethargic, so that we will "waste" less energy with exercise. When ghrelin levels are lower, in contrast, the metabolic rate rises and we are more likely to release some of our stored energy for use, in the knowledge that more supplies are on the way, and we will become more physically active. In this way, ghrelin helps to balance our energy input and output to ensure that we will never run out of fuel.[16]

In the early 2010s Alia Crum and colleagues at Yale and Arizona State University invited participants into the lab on two separate occasions to try out different recipes for a shake.

One was labeled in big letters as "Indulgence: decadence you deserve." Its label encouraged the eater as follows:

> Indulge yourself with this rich and creamy blend of all our premium ingredients—sumptuously smooth ice cream, satin whole milk, and sweet vanilla. It is heaven in a bottle and irresistibly gratifying. Smooth, rich, and delicious!

In the box describing its nutritional information, the label listed the energy content as 620 calories in total (270 of which came from fat). It was accompanied by a picture of a glass cup loaded with ice cream, chocolate sauce, and sprinkles.

The other was called a "Sensi-Shake" for "guilt-free satisfaction."

> Get sensible with the new light, healthy Sensi-Shake. It has all the taste without the guilt—no fat, no added sugar, and only 140 calories. Sensi-Shake is light and tasty enough to enjoy every day.

The illustration for this one was an insipid-looking vanilla flower—the flavor of the drink.

In reality, the shakes on both days were exactly the same, containing 380 calories each. To measure participants' ghrelin responses to these manufactured expectations, Crum's team took blood samples at regular intervals before and after the participants had read the marketing material, and after they had drunk the shakes. For the indulgent shake—"smooth, rich, and delicious"—the ghrelin levels changed exactly as you would hope after a hearty meal, dropping in line with the expected effects on the participants' hunger. For the "sensible," "guilt-free" shake, however, the ghrelin levels barely changed at all.

By creating a change of mindset and with no actual alteration of the shake's nutritional content, Crum's team appeared to have shifted the hormonal profiles of the participants: in one instance, it set them up for greater satiation and an increased metabolism; and in the other, it primed them for greater hunger and reduced metabolism.[17] "When people think they are eating healthily, that's associated with the sense of deprivation," Crum concludes.[18] "And that mindset matters in shaping our physiological response."

The immediate effects of these mindsets can also be seen in the brain regions associated with energy regulation. People taking a low-calorie drink labeled as a "treat" show a more pronounced response in the hypothalamus, for instance, compared with those who had been given a "healthy" drink. Indeed, the response to the "treat" label looked very similar to the activity observed when participants actually drank a high-calorie Ben & Jerry's shake.[19] Based solely on verbal information, the brain was apparently adjusting its predicted energy intake and expenditure—irrespective of the actual content of the glass.

Further research has revealed that our expectations of food can shape everything from the movement of food in the gut to our insulin response. Consider an ingenious study by researchers at Purdue University who wanted to understand why sugary drinks fail to sate our appetite. A bottle of Coca-Cola, after all, has as many calories as a donut, but because of our low expectations, we still feel just as hungry after the drink—and fail to compensate for the energy intake by reducing our consumption later in the day. Like Henry Molaison's ability to eat meal after meal without satisfaction, this phenomenon was hard to explain if you believed that our hunger arises solely from the chemical sensing of nutrients in the gut. But the Purdue researchers suspected that this was due to a widespread assumption that liquids contain fewer calories than foods and are less satisfying—an expectation that directly influences the way they are digested, including the amount of time they are held in the gut.

In one strand of the experiment, participants were given a cherry-flavored drink—but before they drank it, the experimenters gave one of two "demonstrations" of the way the food would react when it met gastric acid in the digestive system. Some were shown the drink mixing into another liquid without changing form, while others were shown the liquid solidifying into a mass—a process that made the nutrient content seem more tangible and substantial. The effects of the manipulation were clear from participants' off-hand comments, which the researchers recorded as they carried out the experiment. Those who believed that the drink remained liquid in the stomach reported feeling very little satiation from what they'd drunk—"It went right through me"—while those who believed that the liquid could transform into a solid felt much fuller. "It feels like I swallowed a rock," one said. "It is very surprising—I feel like I ate a large meal," said another. One participant reported feeling so full, they could barely finish the glass.

As Crum had also found, these reported sensations were reflected in various objective measures. After taking the drink, participants also swallowed a chemical tracker that allowed the scientists to track the course of the drink through the digestive tract. When the participants believed that the liquid had taken a more solid form, it took longer for the liquid to pass from the mouth to the large intestine. This slower transition time could explain why these participants felt fuller for longer. As a result, participants were less likely to snack later, and they consumed around four hundred fewer calories over the course of the day, compared with those who believed that the drink had remained a liquid in their stomach.[20]

In everyday life, the sensory characteristics of a drink—whether it is thick and creamy or thin like water—will influence expectations of satiety. Repeated experiments reveal that the more viscous a liquid is, the more filling we expect it to be, and the more pronounced the physiological response.[21]

Scientists are still coming to grips with the true influence of our expectations on digestion, but an underappreciated study from the 1970s suggests that the effects may even influence our absorption of vitamins and minerals. The scientists in question were examining iron deficiency in Thailand, where it was known to be more prevalent than in many other countries. Their first experiments used a minced-up version of a Thai meal to deliver the nutrient, and they found that the absorption was far lower than might be expected, given the levels of iron content in the food the participants had been given; it just didn't seem feasible that the participants' bodies could take up so little of the nutrient from the meal unless they had a serious health problem, and the resulting lack of iron would have led to much worse anemia than they were really experiencing. This led the researchers to wonder whether the presentation of the meal—in the form of a relatively unappetizing mush—had distorted their results. The food did, after all, resemble the kind of puree we might serve to weaning babies, which is hardly the kind of dish most adults will relish.

To test this hypothesis, the researchers decided to directly compare two forms of the meal—a traditional Thai vegetable curry and a "homogenized" version put through a food processor. The results were astonishing: on average, the participants absorbed 70 percent more iron when the meal was presented in its traditional form, compared to the "homogenized" paste.[22] The team also examined whether the effect would also be present across cultures, so they performed the same test on Swedish participants eating a stereotypically Western meal—a hamburger served with mashed potatoes and green beans. Again, the iron absorption was far higher when the food was served as a recognizable meal, compared with the puree.[23]

In these experiments the presentation of food—and the participants' resulting attitudes toward it—profoundly altered its effects on the body. When food comes in an unfamiliar or unappetizing form, we no longer have expectations of satisfaction and enjoyment, which could have an immediate effect on the release of digestive juices that help us to reap its goodness.

Many people—including self-styled diet gurus—view eating as a purely chemical process, as if we are simply shoveling fuel into a furnace. But these experiments described above show that exactly the same item can be nourishing and satiating, or unfulfilling and nutritionally empty—in large part because of our memories of what we have eaten, our impressions of what it contains, and the meanings that we ascribe to it.

## MEANING-INFUSED MEALS

Looking into the history of these ideas, I was again surprised that mainstream science has taken so long to investigate the role of expectation in diet and nutrition. More than a century ago the Russian scientist Ivan Pavlov found that he could train dogs to associate food with certain cues, such as a horn, a whistle, or a flashing light. (It is apparently a myth that he ever used a bell.[24]) Eventually the cue was enough to get them salivating without food even being present; the release of enzymes in the mouth would kick-start the breakdown of the food into its absorbable nutrients. That's a basic expectation effect—yet few scientists followed up on this work to explore how our broader thinking about food might influence digestion.

Some clues about the effect our brains have on appetite and digestion even lurked in the studies of the placebo effect in medicine. Patients who have been led to believe that they have undergone surgery for obesity—such as stapling the stomach or fitting a gastric balloon—often experience reduced appetite and substantial weight loss, even if they had received a sham treatment; overall they report about 70 percent of the benefits seen by the people receiving the real operation.[25] The research on expected satiety and its physiological effects offers a very natural extension of those findings, yet it took decades for scientists to make the connection. In hindsight, it seems foolish to have ignored the intellectual, emotional, and cultural elements of what we eat, while focusing exclusively on the raw nutritional content of food.

This delay has been a great loss to world health, since an understanding of the role of expectation may offer exciting new tools in our fight against obesity, which currently affects 13 percent of adults worldwide.[26] While many health authorities continue to launch campaigns promoting healthy eating, they have failed to consider the ways that people's beliefs about food and nutrition may be sabotaging their attempts to lose weight.

You can take a test yourself. In each of the following food pairs, does the first item have more, fewer, or roughly the same number of calories as the second item?

| | |
|---|---|
| A regular McDonald's hamburger | 8.5 ounces of grilled ocean cod |
| 1 cup of low-fat yogurt | 2/3 cup of ice cream |
| A banana | 4 Hershey's Kisses |

In reality the calorie content is roughly equal—yet most people believe that the hamburger, ice cream, and Hershey's Kisses contain much more energy than the banana, low-fat yogurt, and grilled cod—over- and underestimating the food's true content by as much as 50 percent. And those errors make a real difference to people's weight; the greater the mismatch in their estimates, the heavier they are.[27]

When researchers examined our associations with different products, meanwhile, they found that people are more likely to link foods like broccoli or salmon with words like "hungry" or "starved"—associations that should reduce expected satiety and increase their hunger at a later point.[28] Surveys, meanwhile, have asked participants to rate statements such as the following. On a scale of 1 (strongly disagree) to 5 (strongly agree), do you think that

- There is usually a trade-off between healthiness and tastiness of food?
- There is no way to make food healthier without sacrificing taste?
- Things that are good for me rarely taste good?

The researchers found a clear correlation between responses to these statements and people's propensity for weight gain: the higher someone scores (that is, the more strongly they agree with these three statements), the greater their body mass index, a measure of weight relative to height and one of the best indicators of unhealthy fat accumulation.[29]

In the past, we might have imagined that these people have low self-control—they just didn't want to give up the momentary pleasure of food—but research on the expectation effect suggests that the truth is more complicated.[30] Imagine you visit a doctor who tells you that you are at risk of obesity. You may respond with good intentions and buy lots of low-calorie foods, but the very idea that they are "healthy"—and all the connotations that word has—will set you up for feelings of deprivation, with direct effects on your physiology. After each meal you may have higher levels of the "hunger hormone" ghrelin running through your body, and your food may even pass through the gut more quickly thanks to the belief that there is less material to digest, meaning that you will feel more ravenous, and your cravings will be more intense. The belief that dieting is inherently difficult will have become a self-fulfilling prophecy. When faced with such difficulties, it is not surprising that even someone with a high level of willpower will struggle to make a lasting change to their eating.

I'll soon explain how we can individually overcome these challenges. The fact remains, however, that our environment is constantly pushing us into making these assumptions, and we need to learn to spot the messages that are creating the deprivation mindset, including food marketing that continues to reinforce the belief that healthy foods are inherently less satisfying. In 2019 Alia Crum analyzed the menus of twenty-six American chain restaurants that offer "healthy eating" options, and she examined the words they use to described different kinds of foods. She found that the entries for the standard offerings were much more likely to include vocabulary that suggests emotions of enjoyment ("crazy," "fun"), vice ("dangerous," "sinful"), and decadence ("bliss," "succulent," "mouth-watering") as well as the sensations of texture ("crispy," "creamy," "gooey") and taste ("tangy," "flavorful")—all of which would suggest a satisfying experience. Health foods, in contrast, were much more likely to contain words evoking simplicity ("plain," "mild"), thinness ("lite," "skinnylicious"), and deprivation ("fat-free," "low-carb"). These descriptions were all about the things the foods *weren't*; in other words, they set up the exact kind of deprivation mindset that will amplify your hunger and send you straight to the cookie jar a few hours later.[31]

Restaurant menus—and food writing more generally—do not have to be like this. As Crum and her colleagues point out, you could easily spice up the descriptions of vegetable dishes, say, with sensual and emotive descriptions that evoke indulgence and enjoyment: "zesty ginger-turmeric sweet potatoes," "sweet sizzlin' green beans and crispy shallots," and "slow-roasted caramelized zucchini bites" in place of "cholesterol-free sweet potatoes," "light 'n' low-carb green beans and shallots," and "lighter-choice zucchini." Not only does that make the vegetables more appetizing at the time, increasing consumption by 29 percent, according to one of Crum's studies; it should also help to ensure that the eater is less likely to snack afterward.[32] Researchers at the University of Bristol have found that simply adding the words "fuller for longer" to a yogurt container significantly increased people's satiety up to three hours later.[33]

As researchers continue to explore the ways our expectations affect digestion, it will be especially important to recognize the importance of factors such as poverty, which can also change the way we perceive certain foods. Low socioeconomic status is a known risk factor for obesity, and there are plenty of potential explanations for this: the relatively high cost of fresh goods compared to convenience foods; a lack of time to prepare nutritious meals; and a lack of access to health care and other support that may guide someone in their weight loss. But recent studies from Singapore suggest that an expectation effect arising from the feeling of financial insecurity may also play a significant role. When people are primed to feel poorer and less secure, they tend to eat sweeter snacks and opt for bigger portion sizes.[34] And this seems to correspond to observable changes in the body's and brain's hormonal response to food.

The participants were first asked to complete an aptitude test, which would ostensibly predict their future career success and income. In reality, the feedback was a sham—each person was told they had scored in the bottom 19 percent, setting them up with the fear that they would struggle in Singapore's competitive society. To highlight and amplify these worries, the researchers presented the participants with a picture of a ladder, which, they were told, represented Singapore's social structure. Their task was to decide where they thought they fell on the ladder, and to compare themselves to the people at the very top. "Think about how the DIFFERENCES BETWEEN YOU might impact what you would talk about, how the interaction is likely to go, and what you and the other person might say to each other," they were told. Once they had finished with their task, the participants were given a milkshake, and blood tests were taken at regular points before and after.

The results were very similar to the effects seen in Crum's original study on food labels—except here it was people's feeling of social and financial insecurity that had set them up with a sense of deprivation, which in turn influenced their hormonal response. The participants who, because of the sham feedback, had placed themselves low on Singapore's social ladder tended to show higher levels of ghrelin when they received the snack, and they felt less full as a result. Their bodies appeared to be primed to start feeding up and storing fat.[35] The participants were debriefed after the experiment, so they should not have suffered any long-term effects, but if you lived with similar feelings of vulnerability for years on end, this altered hormonal response could slowly tip you toward obesity, even if your choices of food were relatively healthy.

In our evolutionary past this would have been a sensible adaptive response to hardship: if we have to worry about our resources in the future, we need to make sure we make the most of what we have today, so it makes sense to eat more while we can and slow our metabolism in order to build up some reserves. Similar responses can be seen among other social animals: those at the lower end of a group's pecking order tend to eat more when the opportunity arises, and they burn energy less quickly, allowing them to build fat stores in case they end up facing shortages in the future. All this would have once protected us in our vulnerability, but in today's "obesogenic" society—where high-calorie food is relatively cheap and readily available—such responses are likely to lead to worse health.

## THE INDULGENCE MINDSET

If we are looking to change our diet, how can we apply these findings? While the new research does not lend itself to any particular diet plan, many regimens involve some form of calorie restriction, and a number of psychological principles can ease that process, curbing cravings while also ensuring that you gain more pleasure and satisfaction from your food.

The most obvious step is to try to avoid liquid calories in sweetened drinks. As we've seen throughout this chapter, the expected satiety of most drinks is very low, meaning that they are unlikely to reduce later cravings. When I am attempting to lose weight, I even try to avoid juices and smoothies, since they will fill me up less than solid food. If you can't live without them, at least try to make them yourself rather than buying an off-the-shelf bottle so that you are more aware of the solids that went into them: the research suggests this small step could have a meaningful effect on your overall satiation.[36]

Be especially wary of high-sugar sports recovery drinks. According to one study, a single shake can contain 1,200 calories—that's around half the average adult's recommended daily calorie intake.[37] In addition to the drink's liquid form, the fact that it is labeled "healthy" will set up the expectation of less satiation, resulting in greater snacking later on.[38] That's not necessarily an issue if you are looking for a quick energy boost to replace calories burned—but it may lead to overcompensation, and if your primary motivation is to lose weight, you might prefer to find a more satisfying way of reviving yourself.

Second, you should maximize your pleasure in the food you eat. It can be tempting when dieting to eat bland, forgettable meals, almost as an act of penitence, but recent research suggests that flavor and texture are especially important during weight loss, as they help to create the sense of indulgence that increases satiety and enhances our hormonal response to food. As a result, I try to look for fiery, spicy options (such as the spaghetti puttanesca mentioned in the introduction to this chapter) and make the most of intensely umami ingredients such as anchovies or Parmesan cheese. The small number of calories that you add to that meal are more than compensated for by the greater satiety that you experience later on, which in turn reduces your subsequent snacking. According to the research on food expectation effects, the very worst thing you can do is to eat something depressingly insipid that leaves you feeling deprived.[39]

Cultivating a sense of indulgence is especially important for times when you enjoy an inevitable treat. Although there may be the temptation to wallow in guilt after you've eaten cake or ice cream, the research shows that you should instead take maximum pleasure in the experience. One snack, after all, should be no reason to throw all your good intentions to one side, and with the right mindset you can make sure that the treat leaves you feeling full, setting your body up to burn the energy you have ingested.

If that seems hard to believe, consider a study that followed 131 dieters over three months. The participants who associated treats such as cake with "guilt" tended to *gain* weight over that period, while participants who associated cake with "celebration" made progress toward their goals.[40] It may be fashionable to label certain foods as "sinful" or "toxic," but psychological research shows that we should avoid such harsh value judgments if we want to make real changes to our eating behavior.

You can amplify these effects by changing the way you frame food before, during, and after eating—starting with a heightened sense of anticipation at what you are about to taste. In a Canadian and French study from 2016, researchers first encouraged participants to vividly imagine the taste, smell, and texture of various sweet treats. The participants were then asked to state their desired portion of a delicious-looking chocolate cake. You might expect that the previous exercise would have increased their lust for the food, leading them to opt for a bigger slice. But most of the participants showed exactly the opposite reaction, opting for a *smaller* portion than those who had not been primed to think about the sensory qualities. By having thought more carefully about the pleasures of eating, they recognized that they could get all the satisfaction they wanted from fewer mouthfuls.[41] The results chime with another experiment, which asked participants to visualize eating either M&Ms or cheese before they were presented with the real product. They subsequently ate substantially smaller portions of the snacks, compared to participants who had imagined another activity.[42]

With a bit of anticipation about what you are eating, it seems that you can make each bite more potent.[43]

Finally, you should avoid distraction while you eat, and be sure to savor each mouthful. It is a cliché, but eating slowly, taking care to chew your food, leaves you feeling more satisfied, since it increases the "orosensory" experience of what you're eating, which can in turn trigger a greater hormonal reaction to the food.[44] And afterward try to make a point of remembering what you have eaten. Whenever you feel tempted to snack mindlessly, think back to your previous meal and try to re-create the memory of eating it; as you remind your prediction machine to incorporate those calories into its projections of energy balance, you may find that you are less hungry than you had assumed.

Don't expect miracles. You can't turn a lettuce leaf into a feast with the power of your mind, and it seems highly unlikely that you will be able to benefit from these small psychological steps while maintaining a severe crash diet. For more moderate regimens, however, these mental shifts could make all the difference to your waistline and—just as important—your mood. Whether you are losing a few pounds or simply maintaining your current weight, you should see each meal as an indulgence and something to be celebrated. This will benefit both your physical and your mental health.

## THE LIFESAVING BENEFITS OF BEING A BON VIVEUR

When applying these principles, eaters in many countries could take inspiration from French culture. While the belief that healthy foods are inherently unsatisfying appears to be relatively widespread in the USA,[45] such attitudes are less pronounced in the UK and Australia,[46] and in France it appears to be more common to hold the exact opposite view. On average, French people are also much more likely to "strongly disagree" with the statement that "things that are good for me rarely taste good," for example, while laboratory studies have shown that labeling a food as "healthy" in France does not reduce satisfaction and pleasure as it does in other Western countries.[47]

Besides thinking more positively about healthier foods, French people also tend to have few negative connotations around treats and desserts. When asked to select a word associated with different foods, such as "ice cream best belongs with *delicious* or *fattening*," French people tend to opt for the more pleasurable alternative, while people in the United States tend to choose the word with more negative connotations. French people are also more likely to endorse statements like "Enjoying food is one of the most important pleasures in my life," compared to people in the United States.

There will be individual variation among individual citizens of any country, of course, and these attitudes may change with time—but overall French people seem to have far more positive expectations about the food they eat and its effects on their body. The effects of these indulgent, celebratory attitudes toward food can be seen in portion sizes and time spent eating. Even when they are indulging in fast food, French people tend to choose smaller servings—since they know that they can gain more pleasure from fewer mouthfuls—and they spend more time eating it, creating a more detailed memory that sets them up with the expectation that they will feel sated for longer.[48] And this appears to make a real difference to their weight. According to the World Health Organization, the average BMI in France is 25.3—lower than in other European countries such as Germany (26.3), and significantly lower than in Australia (27.2), the UK (27.3), and the United States (28.8).[49]

Besides explaining cross-national gaps in BMI, such attitudes can help us to understand puzzling health differences that cannot easily be explained by the nutritional content of the diet itself. The typical French diet contains a higher proportion of saturated fat from butter, cheese, eggs, and cream than a typical English or American diet, yet French people are strikingly less likely to suffer from coronary heart disease than are UK or US citizens. This was once put down to the drinking habits of the French, including the moderate consumption of wine, which contains antioxidant and anti-inflammatory chemicals that help to reduce the tissue damage that comes with age. In reality there are probably many factors, each playing a small role, including each culture's expectations of different foodstuffs and those foodstuffs' effects on our health and well-being.

Remember that people who believe they are more at risk of a heart attack are four times more likely to suffer from heart disease, even when all other factors are considered. In countries like the USA or the UK, messages around food appear to be engendering similar self-fulfilling prophecies. As the authors of one study concluded: "It is not unreasonable to assume that when a major aspect of life becomes a stress and source of substantial worry, as opposed to a pleasure, effects might be seen in both the cardiovascular and immune systems."[50] Thanks to their more positive food culture, however, French people seem less susceptible to this nocebo response. They know that—in moderation—they can have their cake and eat it, too. We would all be healthier for experiencing that same joie de vivre.

### HOW TO THINK ABOUT . . . EATING

- Avoid distractions during meals and pay attention to the food you are consuming. Try to cultivate strong memories of the experience, which will help you to feel and stay sated.
- If you are trying to cut down on snacks, remind yourself what you ate for your last meal. You may find that recollection helps to curb hunger pangs.
- Be aware of food descriptions that create a sense of deprivation. Even if you are looking for low-calorie meals, try to find products that evoke a feeling of indulgence.

- When dieting, pay particular attention to flavor, texture, and presentation—anything that will heighten your enjoyment of the food and leave you feeling more satisfied afterward.
- Avoid sweetened drinks—it is hard for the body to adapt its energy regulation to their high calorie content.
- Enjoy the anticipation of food—this will prime your digestive response and help you to feel more satisfied afterward.
- Don't feel guilty about the occasional treat, but instead relish the moment of pleasure.

## 7

# DE-STRESSING STRESS

## How to turn negative feelings to your advantage

In the late nineteenth century, doctors, politicians, and the clergy began to wage war on a dangerous new vice that was threatening the world's health—and their battle cry can still be heard to this day. The vice was neither opium nor absinthe, but anxiety. As early as 1872 the *British Medical Journal* noted that the "strained and hurried excitement of these times" was exhausting people's nervous energy, leading to mental and physical breakdowns and even a rise in heart disease. "These figures," it noted, "warn us to take a little more care not to kill ourselves for the sake of living." The journal advocated a form of mental "hygiene" that involved stripping unnecessary stress from its readers' lives.[1]

Men and women were frequently prescribed rest cures, and in the United States the perennially anxious could even attend "Don't Worry Clubs," in which members offered mutual support in their abstinence from anxiety. The movement was founded in a small private parlor in New York City by the musician and author Theodore Seward. Americans, he argued, were "slaves to the worrying habit," which was the "enemy which destroys happiness": it needed to be "attacked" with "resolute and persevering effort."[2] Seward went so far as to call the Don't Worry movement an "emancipation," and he relished comparisons with the Prohibition movement that was then gaining ground.[3]

The idea soon caught on, and by the early 1900s the great psychologist William James observed that a kind of "religion of healthy-mindedness" had taken hold. It was accompanied by a "gospel of relaxation," with the aim of turning the mind away from all negative thoughts and feelings, while instead cultivating happiness from within. "Complaints of the weather are getting to be forbidden in many households," he noted, "and more and more people are considering it to be bad form to speak of disagreeable situations."[4] The aim, he said, was to "act and speak as if cheerfulness were already there."[5]

A steady stream of medical research appeared to confirm the dangers of anxiety, and by the 1980s they were considered to be an unquestionable truth, sparking widespread media coverage. At the center of this research was the idea that our evolved stress response, suited to the real danger of predators in the wild, went into overdrive at the slightest challenge—triggering an intense "fight-or-flight" response. "The saber-toothed tiger is long gone, but the modern jungle is no less perilous. The sense of panic over a deadline, a tight plane connection, a reckless driver on one's tail are the new beasts that can set the heart racing, the teeth on edge, the sweat streaming," one *Time* cover story from 1983 declared. "Our mode of life itself, the way we live, is emerging as today's principal cause of illness." Yet again the magazine's readers were advised to take their thoughts into their own hands. "Rule No. 1 is, don't sweat the small stuff. Rule No. 2 is, it's all small stuff," said one cardiologist.[6] The phrase "stressed-out" entered the English lexicon the same year.[7]

From today's media coverage you might assume we are more stressed-out than ever before. We are frequently told that even small repeated stressors—such as the slight irritation of social media feeds—can pose a danger to our physical and mental health, and we're constantly reminded of the best strategies to relieve the burden, from gratitude journals and mindfulness apps to "forest bathing" in nature and expensive digital detox retreats. Whether we like it or not, we are now all members of a global Don't Worry Club.

But what if all the newspaper coverage, and multimillion-selling books, and inspirational speakers, and even some of the scientists themselves have got it wrong? No one wants to feel anxiety if they can help it—but recent research shows that many of our responses to emotions are often a direct result of our beliefs. By demonizing unpleasant but inevitable feelings, we have been creating potent nocebo effects out of otherwise innocuous events. An appreciation of these expectation effects can transform our approach to a whole range of experiences, from burnout to sleeplessness—and may even help us redefine the pursuit of happiness itself.

## THE STRESS CASCADE

To understand the traditional view of anxiety, and why it is wrong, we must first meet a Hungarian-Canadian endocrinologist named Hans Selye, whose pioneering research at the height of the Great Depression provided some of the first clear evidence for the dangers of stress. Like so many great discoveries, Selye's research into the topic began by accident. His primary task had been to identify female sex hormones and to chart their effects on laboratory rats—but he hit a barrier when he found that the rats often fell sick in ways that just didn't make sense given the chemicals he was injecting. He initially feared that his experiments were contaminated until he began to notice that the rats showed a very similar sickness response to a variety of other experiences: if they underwent an operation, if they were placed in the cold or heat, or if they were forced to exercise for too long on their running wheels. How could such different circumstances lead to the same illness? Borrowing a term from mechanics, Selye began to suspect that the general "stress" of all these experiments was making the rats sick, putting them in a state of alarm that eventually ended with exhaustion and illness.

Years of subsequent research detailed the "stress cascade," a kind of physiological chain reaction that creates a state of alarm, and which slowly increases wear and tear on the body. It begins in the brain with two small masses of gray matter called the amygdalae that read input from all the senses and process their emotional content. When the amygdalae identify a threat—such as an oncoming predator—they send signals to the hypothalamus, the same command center that monitors and controls our energy balance and regulates many other elements of the body's physiological state. Eventually the message reaches the adrenal glands, which start pumping out epinephrine, a hormone that has widespread effects on the body.

The most immediate consequences of epinephrine release can be felt in the circulatory system. The heart beats faster, but blood vessels toward the hands, feet, and head constrict—a response that should prevent blood loss if you are injured. Your breathing will become quick and shallow to provide you with oxygen, and you'll experience a sugar rush, as epinephrine releases glucose stored in organs like the liver. To make sure that energy reaches your muscles, the hormone puts digestion and other activities on pause.[8] Your mind, meanwhile, is primed to focus exclusively on the perceived threat and any other dangers within your environment. This is the fight-or-flight response, and it is a perfect adaptation to an immediate threat such as a physical attack.

If the threat subsides—if the predator passes by, for instance—then the epinephrine will ebb away, and you can quickly return to a more restful state. But if you continue to feel danger, then a second wave of hormonal reactions follows, including the release of cortisol, which keeps the brain and body on high alert in the mid- to long term.

It is this maintenance of mental and physiological arousal—over days, weeks, or months—that was thought to lead to the exhaustion and sickness that Selye observed in his laboratory rats, and which was believed to cause disease in humans, too. The racing pulse and constricted blood vessels put extra strain on the cardiovascular system. The continued fluctuations in cortisol damp down the release of beneficial "anabolic" hormones that lead to tissue repair. These prolonged hormonal changes can also contribute to chronic low-level inflammation, which can damage the walls of the arteries and the tissues in the joints. The mind's hypervigilance, meanwhile, reduces overall cognitive performance, since the brain is devoting more resources to coping with the threat rather than considering new and exciting ways of solving problems.

According to Selye, modern stressors such as professional competition, long commutes, and hectic social commitments put us in this chronic state of arousal, and the result, he thought, was an increased vulnerability to a range of illnesses—from arthritis to heart failure—that had started to plague people in industrialized countries. These "diseases of civilization," Selye declared, were "the price we must pay for successful, hard-working people who are subject to mental distress." Selye's investigations into the stress response were so influential that he was nominated for the Nobel Prize in Medicine seventeen times, and many others continued to mine this vein long after his passing in 1982.[9]

Right from the beginning, however, there were reasons to doubt the claims made about stress. The animal subjects of many of the experiments—including those in Selye's initial studies—were placed under extreme strain, generating a kind of blind panic. That's convenient to identify stark physiological changes in the laboratory, but it's not necessarily reflective of the kinds of mild pressures most of us experience. Studies on humans, meanwhile, neglected to consider whether people's expectations might be determining their responses to stress. If we think back to the Don't Worry Clubs in the late nineteenth century, it's clear that our culture has long believed anxiety and nervous tension to be dangerous—particularly the stresses arising from industrialization and urbanization. Because of the mind-body connection, this attitude could shape people's actual responses to challenging events and uncomfortable situations, creating a self-fulfilling prophecy that may have skewed many of the initial scientific findings. If this is correct, it should be possible to shift someone's stress response simply by changing those beliefs.

## GET STRESSED AND GROW

Jeremy Jamieson, a psychologist at the University of Rochester in New York State, has been at the forefront of the scientific research exploring this tantalizing possibility, starting in the late 2000s. His interest in the ways we frame anxiety stemmed from his experiences as a student athlete. He noticed that some teammates would often get pumped up and excited before a game, whereas they would feel nervous and "freak out" before an exam. These were both high-stakes situations—so why was the potential stress of the situation so helpful in one context and detrimental in another?

Jamieson suspected it was due to the way they appraised the different events. On the sports field, the athletes interpreted their jangling nerves as a sign of energy, but in the exam hall, the same sensations were seen as a sign of impending failure. Those expectations could then become self-fulfilling prophecies, shaping the brain's and the body's responses to the stress. In an early experiment to explore this idea, Jamieson recruited sixty students who were planning on sitting the Graduate Record Examinations (a standardized test that is often required to enter graduate schools in the United States and Canada). Before taking a practice exam in the laboratory, half of the participants were provided with the following information:

> People think that feeling anxious while taking a standardized test will make them do poorly on the test. However, recent research suggests that arousal doesn't hurt performance on these tests and can even help performance—people who feel

anxious during a test might actually do better. This means that you shouldn't feel concerned if you do feel anxious while taking today's GRE test. If you find yourself feeling anxious, simply remind yourself that your arousal could be helping you do well.

Not only did this small piece of guidance—a minimal instruction that would have taken less than a minute to read—improve students' scores on the mock exam, but it also helped these participants to perform better on the real test a few months later. The differences were particularly notable in the math section, which is most likely to trigger fear and dread among the exam takers. (So-called "math anxiety" is now considered a real and very common condition.[10]) The average score of the control group was 706, while those who had learned to see anxiety as a source of energy received 770.

It was an astonishing improvement given the brevity of the intervention and could easily sway someone's chances of being admitted into their first-choice university.[11] With a few sentences Jamieson had managed to shift the students' mindsets toward the pumped-up, energized outlook of his sports teammates, and away from the fears that typically debilitated them—with immediate and lasting effects on their performance.

Later studies examined whether a reappraisal of anxiety could alter people's biological responses too, potentially mitigating some of the long-term damage that Selye and others had warned of. In one follow-up, Jamieson examined the effects on the cardiovascular system. Like participants in the first experiment, some of the subjects were reminded that signs of physiological arousal, such as a quickly beating heart or feelings of breathlessness, that we normally associate with anxiety are not necessarily harmful but the body's natural response to challenge, and that increased alertness could actually improve performance. Members of the control group, in contrast, were asked to ignore the feelings and put them "out of their mind" by focusing their attention on a specific point in the room.

Having read these instructions, the participants were then put through a grueling task, known as the Trier Social Stress Test, that is designed to provoke heightened anxiety. They first had to make a short presentation about their strengths and weaknesses—increasing their sense of vulnerability—followed by taking an impromptu mental arithmetic test. To make the task even harder, the people evaluating the participants' performance had been instructed to have negative body language, with crossed arms and frowning faces, meaning that the participants lacked any encouraging feedback that might have quelled their nerves. All the while the scientists monitored the ways each participant's body responded to the anxiety.

The control group showed all the signs you would expect of the classic stress cascade: the heart was racing, but the peripheral vessels were constricted, channeling blood to the body's core. Although they were not in physical danger, they were reacting as if the body were preparing for injury. The people who had reframed their anxious feelings, however, showed a far healthier response. They certainly weren't "relaxed"—the heart was still racing, but it was working more efficiently, with more dilated vessels that allowed blood to flood throughout the body. That's very similar to what happens when we exercise. The stress has energized the body without putting strain on the cardiovascular system.[12] It also allows more blood to reach the brain, providing the cognitive boost that Jamieson had also seen in the GRE results. Distraction didn't work, but reframing did.[13]

Emerging evidence suggests that our expectations may even influence the hormonal stress response. When people are taught that stress can enhance their performance and contribute to personal growth, they tend to show more muted fluctuations in cortisol—just enough to keep them more alert without putting them in a long-lasting state of fear.[14] They also experience a sharper increase in beneficial "anabolic" hormones, such as dehydroepiandrosterone sulfate (DHEAS) and testosterone, which can help to grow and repair the body's tissues; for people who see stress as dangerous or debilitating, there is barely any change.[15] It is the relative ratios of all these hormones that really determine how much wear and tear the body will suffer from a stressful episode—and when people reappraise the effects of stress, they strike a much healthier balance, as if facing an achievable physical challenge rather than a serious existential threat.

Why does reappraisal have this power? For researchers like Jamieson, it all comes down to the brain's predictions, as it weighs up our mental and physical resources against the demands of the task to plan the most appropriate response. If you see your anxiety as debilitating and performance reducing, you reinforce the expectation that you are already at a disadvantage and that you are going to fail—and the brain responds by preparing the body for danger and potential injury. But if you see the racing heartbeat as a sign of energy for an important and potentially rewarding event, you reaffirm the idea that you have everything you need to thrive. "The stress response, instead of becoming this thing to be avoided, actually becomes a resource," said Jamieson. The brain can therefore afford to focus calmly on the task at hand without being alert to every possible threat, while the body can prepare to perform at its maximum capacity, and to potentially grow from the experience, without the risk of being wounded.[16] Afterward the body can return more quickly to all the other useful activities, like digestion, that it normally performs during times of rest.

Besides leading to these physiological changes, attitudes to stress can also transform behavior and perception in quite profound ways. When faced with a difficult challenge, people who see stress as enhancing tend to focus more on the positive elements of a scene (such as the smiling faces in a crowded room) rather than dwelling on potential signs of threat or hostility. They also become more proactive—

deliberately seeking feedback and searching for constructive ways to cope, rather than trying to hide away from the problems at hand. They even demonstrate more creativity. All these changes would mean that they are better equipped to find permanent solutions to the challenges that were causing the distress in the first place.[17]

We now know that our attitudes to stress can have a meaningful impact on all kinds of situations. Reappraising feelings of anxiety has improved people's performance in salary negotiations,[18] for example, while US Navy SEALs with a positive attitude to the stresses of their job showed greater persistence and enhanced performance in training.[19] The shift in mindset has also been proven to improve the experiences of people diagnosed with chronic disorders such as social anxiety, helping them to deal with their fears of social judgment more constructively. Jamieson's team, for example, asked socially anxious individuals to take the Trier Social Stress Test. By reappraising their feelings, the participants were able to give their presentation with fewer visible signs of anxiety; they fidgeted less, showed greater eye contact, and used more open hand gestures and body language.[20]

While many experiments have examined the benefits over relatively short periods, longitudinal studies suggest that these attitudes can also have a significant impact on long-term health. A survey of German doctors and teachers, for instance, found that people's attitudes to anxiety could predict their overall psychological well-being over a yearlong period. Those who saw anxiety as a source of energy—agreeing with statements such as "Feeling somewhat anxious makes me more active in problem solving"—were much less likely to suffer from emotional exhaustion than those who viewed it as a sign of weakness or a threat to their performance.[21]

Our expectations may even override the apparent link between stress and heart disease—one of the most persistent and alarming messages about anxiety. An eight-year longitudinal study of more than 28,000 people, for example, found that high levels of anxiety and mental tension did indeed lead to a 43 percent increase in mortality—but only if the participants believed that it was doing them harm. People who were under high pressure but who believed it to have little effect on their health were actually less likely to die than those who experienced very little stress at all. That was true even when the scientists controlled for a host of other lifestyle factors such as income, education, physical activity, and smoking. Overall, the authors calculate that the belief that stress is harmful leads to the equivalent of around twenty thousand preventable deaths a year in the United States—an astonishing number of people who, like the Hmong immigrants we met in the introduction, are essentially dying from noxious expectations.[22]

As someone who has regularly suffered from anxiety, I was initially a little skeptical of these findings. Often our feelings can hit us like a runaway train—and the idea of overcoming them through a simple reappraisal sounded rather close to the unhelpful and irritating refrain that we should "Get over it." Jamieson, however, emphasizes that the aim is to change your interpretation of anxiety rather than suppressing the feeling itself. That's a vital distinction, since attempts to avoid or ignore our feelings often reinforce uncomfortable emotions and add to their stigma. (Why, after all, would you avoid a feeling if it could be good for you?) With these new reappraisal techniques, you don't need to worry if you still feel breathless and your heart is still racing: the simple point to remember is that those responses are not a sign of weakness and should actually help you to perform at your best and to grow in the future.

Nor does reappraisal require any deception. You're rationally questioning your assumptions about anxiety and reinterpreting the potential effects of your feelings based on substantial scientific research rather than misinformation or unfounded optimism. As we saw with the open-label placebos in Chapter 2, the approaches to pain management in Chapter 3, and the exercise reappraisals in Chapter 5, it's perfectly possible to see beneficial expectation effects without fooling yourself into thinking something that isn't true. Needless to say, the individual benefits will depend on your circumstances. Reappraisal can't make up for a total lack of preparation before an exam or interview. But when you have taken all the practical steps to deal with the situation at hand, the strategy ensures that your feelings are working for you rather than against you.

Many of the existing methods of stress management depend on the power of expectation. We are inundated with apps and books touting the benefits of mindful breathing, for instance. While slow, deep breaths do have some direct physiological effects—they appear to stimulate calmer brain activity, for example—the responses are far greater when people are offered an elaborate description of the apparent benefits.

The same is true of "gratitude journals," which encourage you to note down the things that you value in your life each day. According to numerous magazine articles and websites, this practice is a proven way of buffering the effects of anxiety, and some mental health practitioners have even started to prescribe them as part of their therapy. It is certainly true that the practice improves mood compared to doing nothing at all. Yet a large study published in 2020 found that the effects were less impressive when the gratitude journals were compared to "active control" tasks—such as forming daily "to do" lists, describing the day's schedule, or keeping a record of the day's thoughts (good and bad). This suggests that much of their benefit may arise from the general sense that we are doing something constructive, rather than the specific exercises.[23]

In both cases, the practices leave you feeling that you have more resources to cope with the challenge, which should, in turn, change the way you frame the problem and your anxiety. But if you don't have the expectation that they are doing you good, you may struggle to notice the benefits. The truth is that we all have different associations with a given activity that may boost or undermine its benefits, and if singing in a choir, reading a novel, or playing Tetris leaves you feeling healthier and happier than an hour's yoga practice, you would do far better to embrace this fact than try to suppress your feelings by practicing an activity that leaves you bored and frustrated.

One of the benefits of stress reappraisal is that the potential interventions are incredibly cheap and easy to roll out. A few years ago undergraduates at Stanford University received an email with some logistical information about their first midterm exam for their introductory psychology course. Buried within the message was a paragraph about the potential benefits of anxiety—of the kind that Jamieson used in his first experiment. The small piece of information not only resulted in better results on the midterm exam but also improved the students' overall performance over the whole course.[24]

If you are struggling to imagine the potentially positive effects of anxiety, it might help to identify existing situations in which you already cope well with stress. Perhaps you're like the athletes who inspired Jamieson's original research, and you already understand that the pregame nerves help to pump you up before a game. If so, remembering how you channel your energy on the sports field should help you reframe your nerves before an exam or interview, for instance.

You might also find it useful to frame your anxieties in terms of your broader goals, so that the sensations themselves can be read as a signal that something is meaningful to your life.[25] You're unlikely to feel anxious about something you don't care about in the least. If you feel nervous before a job interview, it's a sign of how much passion you feel for the position that is at stake, and your potential for growth. This way, you stop seeing the difficult situation as a threat (which triggers the fight-or-flight response) and start seeing it as a potential challenge that can be overcome, which makes it easier to reframe nervous feelings as a source of energy that can push you toward success. The benefits appear to build with practice, so be ready to take small steps and allow your confidence to grow over time.[26]

Such an approach helped Billie Jean King, one of history's greatest tennis players, to turn her anxiety to her advantage. As a fifth-grade schoolgirl, she was apparently so shy that she refused to give an oral book report; the mere suggestion of public speaking triggered the classic fight-or-flight response. "The thought of getting up and talking in front of the class absolutely terrified me," she later wrote. "I thought my heart would beat out of my chest and I would die right there." As her tennis career progressed, however, she found a way to reframe those feelings by focusing not on her fears but on the potential for growth that came from difficult challenges. "I realised winning a tournament was the eventual—and desired!—result of all my hard work, and whether I liked it or not, the pressure of public speaking came with the privilege of winning." Her first speech—at a junior tennis tournament—was hesitant, but she managed it without embarrassment or dying from the nerves.

King soon began to see that the same principle—that pressure is privilege—applied to all kinds of situations, and that her anxiety was a sign of her motivation to succeed. "Great moments carry great weight—that is what pressure to perform is all about. And though it can be tough to face that kind of pressure, very few people get the chance to experience it." With that realization she saw that she should embrace rather than suppress feelings of stress—a mindset that allowed her to get through her first Grand Slam wins and the enormous media hype around the Battle of the Sexes match against Bobby Riggs in 1973. As she wrote in her memoir, "At first, I felt obligated to play Riggs, but I chose to embrace as a privilege the pressure that threatened to overwhelm me. This changed my entire mindset and allowed me to deal with the situation more calmly. And as time went on, I began to see the match as something I *got* to do instead of something I *had* to do."[27] The shy fifth-grader who feared that she would die from her nerves during public speaking became one of our greatest athletes and one of sport's most prominent spokespeople.

Ultimately, the use of reappraisal should be seen as one potential tool rather than a "silver bullet"—a useful instrument that can slowly help you to edge out of your comfort zone.

## THE PARADOX OF HAPPINESS

Given the potentially energizing effects of anxiety, we may need to rethink our black-and-white views of many other emotions—and maybe even the "pursuit of happiness" itself. Since the end of the nineteenth century, fears about anxiety have been tied to a more general philosophy of positive thinking—the idea that we should actively cultivate happiness and optimism while "attacking" any negative feeling. This was the "religion of healthy-mindedness" that the psychologist William James described, and it inspired best-selling self-help writers such as Dale Carnegie. The sentiment even hit the charts in 1988 with Bobby McFerrin's song "Don't Worry, Be Happy."

While they are by no means universally accepted, such ideas about the importance of striving for happiness are rife in today's wellness literature. Just consider Elizabeth Gilbert's best-selling memoir *Eat Pray Love*, in which she recounts some advice from her guru: "Happiness is the consequence of personal effort. You fight for it, strive for it, insist upon it, and sometimes even travel around the world looking for it," she writes. "You have to participate relentlessly in the manifestations of your own blessings. And once you have achieved a state of happiness, you must make a mighty effort to keep swimming upward into that happiness forever, to stay afloat on top of it. If you don't, you will leak away your innate contentment."

As Iris Mauss, a psychologist at the University of California, Berkeley, told me: "Wherever you look, you see books about how happiness is good for you, and how you basically should make yourself happier, almost as a duty."[28] And she has spent the last decade showing the ways that message may backfire by adding to the stigma of negative feelings. In 2011, for instance, she asked participants to rate the following statements on a scale of 1 (strongly disagree) to 7 (strongly agree):

- How happy I am at any given moment says a lot about how worthwhile my life is.
- If I don't feel happy, maybe there is something wrong with me.
- I value things in life only to the extent that they influence my personal happiness.
- I would like to be happier than I generally am.
- Feeling happy is extremely important to me.
- I am concerned about my happiness even when I feel happy.
- To have a meaningful life, I need to feel happy most of the time.

This was the "valuing happiness" score—and we can guess that people like Gilbert would rate very high on it. Alongside these beliefs, Mauss also measured participants' subjective well-being: how content they rated themselves to be at that time; the number of depressive symptoms they exhibited; and the ratio of positive to negative emotions they were experiencing (the so-called "hedonic balance").

Contrary to the claims of so many inspirational speakers and writers, Mauss found that the people who valued happiness the most, and strove the hardest to achieve it, were unhappier on *every measure* that she considered. Following the "religion of healthy-mindedness" and striving to cultivate good feelings in every moment would be just about the worst thing you could do for your well-being.

In a second experiment, Mauss asked half her participants to read some text about the importance of happiness—the kind of material that is found in many newspapers or magazines. They then watched a heartwarming film about a figure skater who had won an Olympic gold. Once more, the results were highly counterintuitive: rather than savoring the joy of the story, the participants who had read the article about happiness were much *less* likely to be moved by the clip, compared to those who had not read the article. They focused so much on how they *should* feel that they were ultimately dissatisfied when the clip didn't bring the joy they'd expected.[29] The harder we try to be happy, the less happy we are, partly due to a heightened sense of self-consciousness that makes it hard to fully appreciate small, spontaneous pleasure.

Equally important are the ways that a constant fixation on happiness can lead us to frame our negative feelings and the small, inevitable upsets that are part of life as something inherently undesirable and damaging. To test this possibility, scientists primed people to think about happiness by seating them in a room filled with motivational posters and books on well-being before they completed a frustratingly difficult test. These participants subsequently ruminated far more about their failure to find the right answers compared to participants who had not been encouraged to think about the benefits of positive feelings.[30]

The more you stigmatize a feeling, it seems, the more likely you are to dwell on that emotion when it does finally enter your life—tipping your hedonic balance from positive to negative and making it much harder to recover from the emotional blow.

You can see whether you fall into this trap yourself. On a scale of 1 (never/very rarely true) to 7 (very often/always true), how would you rate the following statements?

- I tell myself I shouldn't be feeling the way that I'm feeling.
- I criticize myself for having irrational or inappropriate emotions.
- When I have distressing thoughts or images, I judge myself as good or bad depending on what the thought or image is about.

- I think some of my emotions are bad or inappropriate and I shouldn't feel them.
- I believe some of my thoughts are abnormal or bad, and I shouldn't think that way.

In a study of around one thousand participants, Mauss found that the higher people scored on this questionnaire, the more likely they were to report symptoms of depression and anxiety, and the worse they scored on general measures of life satisfaction and psychological well-being. People who reported accepting their thoughts and feelings, in contrast, without characterizing them as "bad" or "inappropriate," tended to have better psychological health.[31]

Exactly the same patterns were evident in a German survey published in 2016, which found that people who see meaning in unpleasant feelings tend to be far happier than those who would rather eliminate them. The researchers, based at the Max Planck Institute for Human Development, asked participants to rate various emotions, such as nervousness, anger, or feeling downcast, on four dimensions: their unpleasantness, their appropriateness, their utility, and their meaningfulness. Disappointment, for example, may feel unpleasant—but you could recognize that it is a necessary means of processing a failure and learning from previous mistakes, leading you to rate it highly for appropriateness, utility, and meaningfulness.

Just as Mauss's studies predicted, the participants who interpreted their feelings in this way tended to fare much better on measures of mental and physical well-being, including their risk of illnesses such as diabetes or cardiovascular disease and even their muscle strength (which was considered a general indicator of fitness). Indeed, the capacity to see value in unpleasant emotions almost completely eliminated any link between their health and the *actual* number of upsets the person reported experiencing. Even if the participants reported feeling distressed at multiple points during the three-week study period, the act of accepting and assigning positive meanings to those experiences helped them to recover more quickly, without leaving a permanent mark on their physical and psychological well-being.[32]

For an example of how you might apply new meanings to an uncomfortable emotion, imagine you have had a disagreement with your boss, in which they unfairly scolded you for your lack of progress on an important task. How would your anger influence your performance for the rest of the day? You might think that it would result in nervous agitation, distraction, and impulsivity, shredding your concentration, or you could believe that the feelings of anger will increase your determination and resolve. Either set of expectations can make a noticeable difference to your actual behavior, as Maya Tamir at the Hebrew University of Jerusalem has demonstrated.

Tamir first asked her participants to listen to various musical tracks—a technique that is commonly used to prime people's moods in the lab. Some listened to the finale from the sound track of a horror film (*The Curse of the Werewolf*) and two tracks by symphonic metal band Apocalyptica, all of which were designed to get them feeling slightly angry, while others listened to more relaxing ambient music. The researchers then paired each participant with a partner and asked them to complete a simple negotiation game. The pair was presented with a pile of different-colored chips that were assigned a monetary value, and then asked to agree on a way to divide the chips among themselves as best they could. As an incentive to perform well, the participants were told they could keep whatever money they made. To make the negotiation harder, however, the value associated with each color was different for each person—something that was good for one side might not be for the other. In this way, the study mimicked the same kind of wrangling that is experienced in a divorce proceeding, in which different objects may be more or less desirable to each party.

Just before the task, people were also given some friendly advice, reportedly from previous participants. Some were told: "I think the most important part of the whole process is figuring out how to act so you can get the most money for yourself. Throughout the negotiation, I was persistent. Eventually I was reasonable, and my partner gave me what I wanted." Others were told: "I think the most important part of the whole process is figuring out how to act so you can get the most money for yourself. Throughout the negotiation, I was persistent. Eventually I got angry, and my partner felt compelled to give me what I wanted."

The participants' behavior mirrored that of the subjects of Jamieson's studies of anxiety. When they were told their anger could be useful, they successfully turned their frustrations to their advantage and performed significantly better than did the calmer participants.

To confirm the expectation effect, Tamir conducted a second study, using an action-based computer game requiring fine motor skills. Once again, expectations shaped the way that people's emotions influenced their performance. Angry participants killed about twice as many enemies in the game when they were told anger was useful, compared to when they were told that a cool head was necessary to prevail. Overall, the angry participants were about three times better than the calmer participants—provided they knew about the benefits of that emotion and its potential use as a source of energy.[33]

People with high emotional intelligence already have expectations of anger's benefits, Tamir has shown, and so may many athletes. Frustrated ice hockey players tend to be more accurate in penalty shoot-outs than those in a calmer frame of mind, for example, while basketball players take more accurate shots when they believe they've been wronged.[34]

Clearly, it will take more than an expectation effect to solve severe anger management issues, but Tamir's and Mauss's work underlines that many other negative emotions, besides anxiety, may be a product of our expectations. We don't need to *enjoy* such feelings—but recognizing their potential value will allow us to channel them more effectively, and to recover more quickly once they have served their purpose. By accepting the bad with the good we can begin to resolve the paradox of happiness.

## THE COMPLAINING GOOD SLEEPER

"How do people go to sleep?" Dorothy Parker asked in her story "The Little Hours." "I'm afraid I've lost the knack." Anyone who has suffered short- or long-term insomnia can empathize with the narrator's struggles—including her thoughts of "busting myself smartly over the temple with the night-light." As strange as it may sound, difficulties with sleep and their effects on our health and well-being often bear a striking resemblance to waking stress responses.

For one thing, insomnia is often fueled by the same ruminative, catastrophizing thought process that amplifies anxiety and decreases happiness.[35] The more you fear not falling asleep, the more the mind begins to race just before bed, and the harder it is to actually drop off—as Parker noted in her story. This might explain why the placebo effect can account for about 50 percent of the success of sleeping pills: the expectation that they will bring relief helps to cut through rumination.

People's concerns about sleep will lead them to underestimate how much sleep they actually are getting, and erroneous beliefs about sleep loss will become a serious source of worry, setting off a vicious cycle. The prediction machine will decide that we are ill equipped to deal with the day's challenges, meaning that everything starts to feel more stressful—with the accompanying physiological effects.

As proof of this expectation effect, various experiments have compared more objective measures of sleep, such as recordings of nocturnal brain activity, with the participants' subjective opinions of how much sleep they felt they *should* be getting. Amazingly, the two facets are not very closely linked. Around 10 percent of people are "complaining good sleepers" who believe they are constantly sleep deprived, even though they are actually getting enough shut-eye. A further 16 percent are "non-complaining bad sleepers," who—for various reasons—fail to obtain the recommended seven hours of unconsciousness a night but don't feel anxious about their lack of sleep. And it's the complaining good sleepers who are more likely to suffer from symptoms such as poor concentration, fatigue, depression, anxiety, and suicidal ideation, while non-complaining bad sleepers are remarkably free of ill effects. Even the objective physiological consequences of insomnia seem to depend on expectations; insomnia has been found to raise blood pressure, for example—but this only occurs among the "complaining bad sleepers."[36] (The healthiest people, of course, are those who have good sleep and a positive view of their slumbers.)

To further test the effects of our sleep beliefs, teams of scientists in Colorado and Oxford provided a set of people with sham feedback about their sleep quality—essentially creating a sample of complaining good sleepers. The following day, the scientists asked the participants to complete tests of memory and attention. To test numerical processing, people were asked to listen to a string of numbers, spaced 1.6 seconds apart, and add up the last two digits each time they heard a new number; to test their verbal fluency, they had to produce as many words as possible beginning with a certain letter.

In each case, the participants performed exactly as if the sham feedback had been real. If they believed their sleep was poor—like the complaining good sleepers—then they struggled to complete the mental arithmetic and vocabulary tests; if they believed their sleep was better than average, their mental abilities were much sharper. The negative expectations also resulted in heightened feelings of fatigue and low mood.[37]

This expectation effect is so strong that the author of one meta-analysis concluded that "worry about poor sleep is a stronger pathogen than poor sleep."[38] A recognition of this fact should change the way we treat sleeplessness. According to the CDC, around 8 percent of the adult population in the United States regularly takes medication to help them sleep—amounting to around seventeen million people in total.[39] But according to the work on complaining good sleepers, around 40 percent of those people have no objective problem with their sleep—and may benefit from breaking the cycle of maladaptive thoughts that are leading to their daytime symptoms.

One of the easiest ways to do this is to adopt a more accepting attitude to feelings of restlessness without thinking too much about the consequences for the following day. (Some studies have even found that deliberately attempting to stay awake can, ironically, cure people's insomnia by removing the sense of struggle that usually comes with sleeplessness, though it is easy to see how that approach could backfire over time.) Sure enough, people who are asked to passively monitor their thoughts and feelings, without actively fighting them, take significantly less time to fall asleep.[40]

We can also attempt to reappraise specific assumptions about sleep so that we acknowledge the importance of a good night's rest without seeing moderate sleep loss as a catastrophe. Along these lines, psychologists studying insomnia have compiled a list of "dysfunctional beliefs and attitudes about sleep," all of which promote an overly pessimistic view of insomnia. They include the following:

- Misconceptions about the causes of insomnia ("I believe insomnia is basically the result of aging and there isn't much that can be done about this problem," or "I believe insomnia is essentially the result of a chemical imbalance")
- Diminished perception of control and predictability of sleep ("When I sleep poorly on one night, I know it will disturb my sleep schedule for the whole week")

- Unrealistic sleep expectations ("I need to catch up on any sleep loss")
- Misattribution or amplification of the consequences of insomnia ("Insomnia is destroying my life," or "I cannot function without a good night's sleep")
- Faulty beliefs about sleep-promoting practices ("When I have trouble sleeping, I should stay in bed and try harder")

None of these beliefs have a strong factual basis: as the research on "non-complaining bad sleepers" shows, we are actually much more resilient to moderate sleep loss than people assume. And people who are taught to confront and question these expectations enjoy greater overall sleep quality, relieved daytime fatigue, and fewer depressive symptoms.[41] The trick is to take it slowly rather than hoping for immediate, total relief, and you don't need to tackle every issue at once. You might start out by observing whether you dropped off a little more quickly than expected, or whether you managed to achieve a little more sleep than you anticipated after a night's disrupted sleep, for example, and then build on those small gains. With time, you should find that the "knack" of falling asleep and awaking refreshed has returned.

## EUSTRESS

Our understanding of the ways that our expectations shape the biological reality of our emotions has only just started to flourish—but it's not too early to reap its fruits. Whether we are agonizing about the working day or tossing and turning in frustration at night, our interpretations of our feelings can do more harm than the feelings themselves. Often just a simple reassessment of our assumptions enables us to thrive.

"People need to be reminded of how strong the influence of our brain is on our body," says Jeremy Jamieson. It's not as if the components of the stress circuit have their own sensory organs to automatically gauge a danger, he says—we are always responding to a complex mental construct formed from our beliefs and expectations. And it's now within our power to change that construct. "We can tell it what to do—and we do that through these appraisal processes," he says. It may not be a panacea for every stress we face, but I've certainly found that quick reappraisal can be one useful tool for coping with everyday anxieties that would once have left me feeling unhappy and exhausted.

Hans Selye—the father of stress research—started to come to these conclusions in his last decade. He had, after all, lived a busy and hectic life, with research, writing, and relentless international lecture tours, but he thrived on the constant stream of challenges. And so, after four decades of describing the dangers of stress, he began to suspect that our attitudes might play a role in our responses. As he noted in 1977 in his autobiography, a kiss from a lover could bring about many of the same changes—the rushing heart, the breathlessness—that come from fear. The only difference is the interpretation. Selye even coined a term, "eustress," to describe the energizing, beneficial feelings that can come from taking on new challenges, and he argued that life would be meaningless without them. Stress, he concluded, "is not what happens to you, but the way you take it."[42]

With our advanced understanding of expectation effects, we can finally put that sentiment into action—a truce, perhaps, in the century-old war on worry.

### HOW TO THINK ABOUT . . . STRESS, HAPPINESS, AND SLEEP

- Try to adopt an accepting attitude to unpleasant feelings, rather than actively suppressing them.

- When dealing with anxiety, consider the potential benefits of physical feelings. Fast breathing and a racing heart, for instance, help to transmit oxygen and glucose to your body and brain, providing you with the energy you need to cope with the challenge, while sweating helps cool the body as it works hard to meet its goals.
- Can you relabel your feelings? Anxiety can feel much like excitement, for example, and reminding yourself of these similarities can help you to feel more energized.
- If you have an imaginative mind, visualizing the ways that anxiety could boost your performance in specific situations can cement the message, leading to longer-term effects.
- Regularly reinforce your knowledge of these expectation effects. If you often feel stressed in the workplace, it might help to have notes or posters outlining the principles of this chapter around your desk, or to place reminders in an online calendar.
- If you struggle to fall asleep, try to accept the feeling of restlessness without judging it, and remind yourself that you will still be able to function the next day, even if your sleep has been less than optimal.
- Try to get more objective data on your sleeping habits—using, for instance, a phone app or sleep monitoring device—and consider whether you may be a "complaining good sleeper." If so, question your assumptions using the "dysfunctional beliefs and attitudes about sleep" scale on pages 176–77.

# 8

# LIMITLESS WILLPOWER

## How to build endless reserves of self-control and mental focus

If you were watching closely during Barack Obama's presidency, you might have noticed that he wore the same style of blue or gray suit for almost all public occasions. This was not a fashion statement but a life hack: he thought that he could save his mental reserves for the responsibilities of the presidency if he avoided having to make small but irrelevant decisions that would sap his concentration.

Michelle would apparently roll her eyes at his behavior. "My wife makes fun of how routinized I've become," he said.[1] But Obama is not alone in this energy-saving scheme. Arianna Huffington, Steve Jobs, Richard Branson, and Mark Zuckerberg are all said to have simplified their wardrobe as a way of preserving their brains for loftier tasks. "I'd feel I'm not doing my job if I spent any of my energy on things that are silly or frivolous about my life," Zuckerberg told one interviewer.[2]

Their reasoning was apparently based on solid scientific research, which Obama cited in a *Vanity Fair* interview. For decades, researchers have assumed that any kind of mental effort—such as making decisions, avoiding distraction, or resisting temptation—draws on the brain's stores of glucose. When we wake up in the morning, we have that vital fuel in abundance, but it dwindles every time we exercise our mind, leading to lapses in concentration and self-control as the day progresses. As one expert told the *Financial Times*: "There is only so much quality thinking your brain can do each day."[3] According to this theory, the limits of our mental reserve could explain procrastination at work. Every moment that we exercise self-control by concentrating on the task at hand, it saps a little of our energy, until we eventually can't resist the urge to take a look at Facebook, Twitter, or YouTube. And so we waste hours in the midafternoon lull as the clock counts down to the end of the day.

Importantly, the same mental reserves are believed to power many different tasks, meaning that effort in one area can supposedly lead to lapses in another. This is said to be the reason that some people feast on junk food following a hard day in the office: after a long time concentrating, we lose the strength required to resist the temptation of the cookie jar. Late-night splurges on Amazon and eBay could similarly be blamed on a fatigued brain: when our mental reserves are already down, we can't resist spending our money on useless goods we later regret. Some authors have even claimed that our limited mental reserves could explain why high-powered people cheat on their spouses. With a relentless work schedule sapping his decision-making energy, someone like Bill Clinton would have been almost helpless to control himself, according to this theory.

Does it have to be this way? The idea that our mental resources are limited and can become exhausted over time certainly chimes with many of our own experiences at home and at work. We don't have to have a degree in psychology, after all, to describe someone as having a "short fuse," to think of our patience "wearing thin," or to argue that you "can't burn the candle at both ends." And that is precisely the problem with so much of the research on willpower: like studies on stress, it failed to question whether we simply place these limits on ourselves through an extremely prevalent expectation effect.

The truth is that most of us are tapping only a small part of our potential while a huge reservoir is just waiting to be released. Intriguingly, some cultures already have the widespread view that mental focus and self-control increase with effort, and this is reflected in people's behavior.

What we once assumed to be a biological limit is really a cultural artifact, and by learning to change our expectations we can make better use of the brain's enormous reserves. This understanding can even help us to recognize the true power of superstition and prayer—for believers and atheists alike.

## THE DEPLETED EGO

The prevailing theories of mental focus and self-control—which are often lumped together under the term "willpower"—can be traced to the father of psychoanalysis, Sigmund Freud. He saw our psyche as having three components: the id, the ego, and the superego. The id was unruly and impulsive, while the superego was chaste and censorious, dictating the most moral or socially appropriate course of action. The pragmatic ego stood between these two warring entities, reining in the id on the instructions of the superego. But it needed energy to decide what was best for us and to take the correct course of action; otherwise our baser impulses would get the upper hand.[4]

It was only in the late 1990s that psychologists came to test Freud's theory systematically, with Roy Baumeister leading the way. In his first experiment, the psychologist recruited students under the guise of a taste perception test. Arriving in the lab, the students found two bowls on the table—one full of radishes, and one full of cookies. As the experimenter left the room, the luckier participants were told to feast on two or three cookies; the unluckier ones were told to taste the radishes without taking a cookie. (Unbeknownst to the participants, the researchers were hidden behind a two-way mirror to make sure they didn't cheat.) After they had finished the tasting, the participants were given a complex geometric task—so complex, in fact, that it was actually impossible to solve. If the participants wanted to stop, they were told that they could ring a bell, and the scientists would come and collect them. Otherwise, they were given thirty minutes to wrestle with the problem.

Baumeister predicted that the effort of tamping down the temptation to eat the cookies would exhaust the participants' reserves of mental energy, so that the radish-eaters would have less resilience on the problem-solving task. And that was exactly what he found. On average, the people in the radish group lasted around 8.5 minutes at the task before they gave up and rang the bell, compared to 19 minutes for the participants who had been free to eat the cookies—a huge difference in mental stamina.[5]

Based on these findings, Baumeister saw our willpower as a mental muscle that tires with time. As a homage to Freud's theories, he described the mental exhaustion that comes from exercising self-control and focus as "ego depletion," and further research soon offered hundreds more examples to support the theory.[6] After subjects had been asked to watch a Robin Williams clip without laughing or smiling, for instance, they were then less able to focus on solving a series of anagrams.[7] Or participants had to ignore annoying messages popping up on screen while listening to an interview—and they became more distractible on tests of logic and reading comprehension as a result of their depleted mental resources.[8]

In one of the studies that apparently inspired Obama to pare down his own decision-making, students were asked to pick courses for their degree—a choice with serious potential ramifications for their academic success. Because of the mental depletion that came from focusing on this decision, they were subsequently more likely to procrastinate rather than study for a potentially important math test.[9] Studies on consumer behavior, meanwhile, showed that depleted participants who had been forced to remain focused and engaged while reading aloud the boring biographies of scientists were subsequently more likely to make impulse purchases.[10]

In each case, practicing self-discipline and mental focus in one domain—whether resisting temptation, avoiding distraction, solving a difficult problem, planning for the future, or exercising emotional restraint—led to worse performance in another. The evidence appeared to be concrete: we can control our mind and behavior for only so long before we begin to tire.

These experiments were particularly striking because the "depleting" exercises were not especially taxing in the grand scheme of things, suggesting that many of the small trials that we face every day might have similar effects. "When you . . . hold your tongue, resist an urge to smoke, drink or eat, restrain aggression, postpone using the toilet, feign mirth at an inane joke, push yourself to keep working, it depletes some crucial energy and leaves you with less available for meeting the next challenge," Baumeister wrote in *The Psychologist* magazine in 2012.[11] Indeed, when Baumeister gave participants a phone app prompting them to record their thoughts at random intervals throughout the day, he found that the average person spends about a quarter of their day resisting desires, from sex to time on social media.[12] With so many demands draining our resources, it's not surprising that so many of us find it hard to maintain our willpower from time to time.

Brain scans were even able to identify a pair of brain regions—the prefrontal and anterior cingulate cortices—that appeared to be involved in all these different forms of ego control. But to cement their theory, the researchers really needed to be able to identify the fuel that became depleted over time. They settled on the molecule glucose, which also powers our muscles, as the prime candidate.[13]

Exercising our focus and willpower can, after all, feel akin to a physical ordeal. The novelist Edward St. Aubyn apparently has to be wrapped in a towel as he writes, since he sweats so profusely while creating his novels[14]—as if the intense concentration were taxing his whole body. And the research shows that the effort of self-control can indeed lead to signs of physical strain, such as increased perspiration.[15]

In support of the glucose theory, Baumeister points to studies using positron-emission tomography (PET) that can measure energy consumption in the brain, and which show an increase in glucose metabolism in the frontal regions of the brain as we engage in effortful tasks.[16] Baumeister's experiments also found an apparent correlation between people's glucose levels and the effects of ego depletion: the lower their blood sugar, the lower their willpower. Even more compellingly, his team found that delivering a quick sugar rush via a glass of lemonade restored participants' mental focus and self-control when they were feeling depleted.[17]

The apparently rapid exhaustion of our brain's resources—often after just five minutes of mental activity!—may sound like bad news, but Baumeister's research nevertheless offered lots of practical suggestions to make the most of our limited reserves. He found that self-control and mental focus can become stronger with practice, like working a muscle—and as we saw with the effects of ego depletion, the benefits could spill over into multiple domains. Participants in one of the first experiments were encouraged to correct their bad posture over a two-week period, and they subsequently showed greater perseverance on a laboratory test. People who committed to avoiding snacks, meanwhile, were subsequently twice as likely to give up smoking, while those who attempted to curb swearing in their everyday speech were better able to stay patient with their romantic partners.[18] Somehow the brain was learning how to expand its reserves and cope with depletion.

Baumeister argued that the surest way to boost overall self-control is to change your environment, avoiding small, everyday ordeals that slowly sap your energy so you can spend it on the things that really matter. If sweets are your particular vice, you might avoid having them in your home or on your desk at work, so that you aren't exhausted by the continual temptation to open the cookie jar. Or if you are easily distracted by your phone during work, you might put it away in your locker. And if you are the president of the United States, you might simplify your wardrobe and diet to eliminate unnecessary choices and preserve your decision-making energy for matters of national importance.

It all seemed so certain.[19] Yet some recent studies have questioned whether ego depletion is really as inevitable as we once thought. Baumeister's theory, it seems, was missing something essential: the power of belief to control the brain's resources.

## THE MIND'S THE LIMIT

Experience in any profession tells us that some people find the same mental effort much more tiring than others. Just think of the people you know. While some are exhausted at the end of the working day, others seem to have boundless reserves that allow them to read hundreds of novels, play in an orchestra, or write a screenplay. These individual differences may partly depend on our beliefs about the tasks themselves. You might have been brought up to think that reading is hard work, whereas playing music is a form of relaxation, or vice versa, and those beliefs will determine how tiring you find the respective activities. That's worth bearing in mind if you're a parent, teacher, or manager giving instructions to other people. Dutch researchers have shown that simply being told that we may find an exercise to be energizing, rather than fatiguing, can reduce the sense of depletion, so that participants are more persistent and focused—so don't overemphasize the difficulty of a task before someone has tried it for themselves.[20]

Even more powerful, however, are our expectations about our own capabilities and our reactions to hard mental work *in general,* according to some game-changing work by Veronika Job at the University of Vienna in Austria. She has shown that our beliefs about the brain's resources—whether we see them as finite or non-limited—can powerfully change our experience of ego depletion and our capacity to remain self-controlled and focused under pressure.

Working with researchers at Stanford University in the late 2000s,

Job first created a questionnaire to test her participants' "implicit theories" of concentration and self-control, with a series of statements that they had to score on a scale of 1 (strongly agree) to 6 (strongly disagree). They included the following:

- When situations accumulate that challenge you with temptations, it gets more and more difficult to resist the temptations.
- Strenuous mental activity exhausts your resources, which you need to refuel afterward (e.g., through taking breaks, doing nothing, watching television, eating snacks).
- When you have completed a strenuous mental activity, you cannot start another activity immediately with the same concentration because you have to recover your mental energy again.

And

- When you have been working on a strenuous mental task, you feel energized, and you are able to immediately start with another demanding activity.
- If you have just resisted a strong temptation, you feel strengthened, and you can withstand any new temptations.
- Your mental stamina fuels itself. Even after strenuous mental exertion, you can continue doing more of it.

The people who agreed more with the first set of statements were considered to have a "limited" theory of the mind's resources, while those who agreed with the second were said to have a "non-limited" theory of the mind's resources. (In the actual experiment, the statements appeared in a single combined list, along with "decoy" statements that were added to avoid arousing the participants' suspicion about the study's purpose.)

After rating the statements, the participants were given an arduous exercise, in which they had to cross out certain letters in every word on a typewritten page: a boring but fiddly task that was deliberately designed to "deplete" their resources. Finally, they took the Stroop test, a standard test of concentration, in which different color words appear in different colored text. The participants' task is to report the color of the letters, irrespective of the word that is presented. (You might see the word "red" appearing in blue, "black" appearing in orange, or "yellow" appearing in yellow. The correct answers would be blue, orange, and yellow.)

If all this sounds slightly tiresome, you can imagine how some of the participants felt. And the people with the "limited" mindset reacted to the tasks in exactly the way you would predict from ego depletion theory: the proofreading task exhausted their minds, causing them to lose concentration on the Stroop test. As a result, their accuracy was much worse than that of a control group who took the Stroop test without having to complete the text correction.

The people with a non-limited view of the mind, in contrast, showed no signs of fatigue after the first exercise. They actually performed just as well as the control group who had taken the Stroop test when fresh without the dull but tiring proofreading task. Astonishingly, Job's results seemed to show that the consequences of ego depletion are real—but *only if you believe in it.*

Job next recruited a new set of participants to see if she could sway their beliefs, and whether that intervention would alter their performance. Instead of seeing the full questionnaire, half of the subjects were shown the "limited" statements given above, while the rest were given the "non-limited" statements—a subtle intervention that was meant to prime one or the other mindset. The participants were then given the text-correction task and the Stroop test. And the effects were huge: those who had been exposed to the idea that concentration can build with effort were around twice as accurate on the Stroop test, compared to the people who had been primed to think that their resources would become depleted.[21] This proved causality: through nothing more than a nudge toward one belief or another, Job had strengthened or weakened their willpower. Indeed, the people who had been primed with the non-limited views actually performed *better* if they had done the so-called depleting task than if they hadn't exerted themselves before the Stroop test. Their beliefs that mental effort could be energizing had become a reality.

When I spoke to Job about this work, she told me that when she first presented her ideas at a colloquium in her native Switzerland, other researchers were skeptical of the possibility that expectations could have these effects. But these game-changing findings have since been replicated many times—with evidence of significant long-term effects. Using participants' diaries of their day-to-day activities, for example, Job found that people with the "non-limited" view of their mind's resources were better able to recover after a long and tiring day, with higher expectations about what they were going to achieve the following morning, and they achieved greater productivity as a result. Amazingly, she found that people with the non-limited view of the mind were actually *more productive* (than their own average day) after a particularly demanding day compared to a less demanding day. Far from exhausting them, the extra difficulties had boosted their stamina and galvanized their motivation to meet their goals.[22]

The effects of these mindsets are especially salient during the most stressful times, such as the run-up to exams. Those with the limited mindset report greater fatigue throughout the period, which leads to greater procrastination, and students' grades and emotional well-being suffer as a consequence. Because of their depleted self-control, they are also more likely to feast on junk food and resort to impulsive spending to lift their mood—the classic signs of ego depletion. Those with the non-limited view, in contrast, find it much easier to maintain their studies and achieve good grades without procrastinating or neglecting their health.[23]

Our beliefs about willpower can even determine our responses to chronic illness. Job has studied a cohort of people with diabetes, revealing that their mindsets influenced how likely they were to follow doctors' orders. Overall, people with the non-limited view of their willpower were more diligent in acts of self-care (such as keeping a record of their blood sugar levels), taking their medication, and keeping their weight under control.[24] If you feel mentally exhausted at the end of the day, you may be less inclined to take care of yourself, but people with the non-limited view of the mind did not fall into this trap and were healthier as a consequence.

The limited view of willpower may be far more common in the West, but this attitude is not universal. Working with Krishna Savani at Nanyang Technological University in Singapore, Job has shown that the non-limited view of the human mind is much more common among Indian students than among people in the United States or Switzerland—and their mental stamina is much greater as a result.[25]

Job and Savani argue that the greater prevalence of the non-limited belief in India may arise from various religious traditions, including Buddhism, Hinduism, and Jainism: adherents practice mentally taxing activities that are explicitly designed to boost concentration and self-control. They point to the yogic practice of trataka, which involves concentrating your vision on a single point—such as a black dot or the tip of a candle flame—while ignoring all other distractions. Trataka is essentially the kind of attention task that would have been used by Western scientists to deplete our resources. For practitioners of yoga, however, it is seen as a way of "cleansing" the mind, preparing it for further concentration, and regular repetition of the exercise appears to cement the idea that focused mental effort can be energizing rather than fatiguing—leading to greater concentration and self-control in many areas of life.[26] It is interesting to contemplate how different the scientific understanding of willpower could have been if the first experiments on mental fatigue had been conducted in a non-Western culture.

## THE UNIFIED THEORY OF WILLPOWER

These findings may seem to ring the death knell for ego depletion theory. There is, however, a way of reconciling Baumeister's and Job's theories if you think about the way the brain is managing its energy levels. According to one view, the prediction machine is acting like an accountant, parceling out our resources so that we don't run our supplies of glucose (and whatever other sources of mental fuel) dangerously low. As we saw in Chapters 5 and 6, the body's sensors cannot judge our energy intake or expenditure very accurately. This means our inner accountant can be swayed by our expectations, including our beliefs about willpower.

If you see your resources as limited, it makes sense for your brain to operate stingily, reducing its consumption of glucose after an effortful activity. That way, it will eke out its remaining energy reserves and avoid exhausting them before you have a chance to replenish them. In such cases, the sense of depletion isn't just imagined. The brain really is reducing its own energy usage, a physiological consequence of your expectations—in just the same way we might all tighten our belts and reduce our spending as we await the next paycheck.

If you think that you have unlimited resources, however, the internal accountant is much less of a miser, and it releases whatever supplies you need, since it no longer fears running out. Your brain uses as much fuel as it needs in the belief that more will be available, meaning that you can maintain a consistent effort while studying, resisting temptation, or making tough decisions. It has no need to cut its energy expenditure and reduce its performance.

If true, this might explain why willpower grows with practice: the exercises that Baumeister and others had set their participants helped them to prove to themselves that their mental resources were less easily depleted than they once thought, allowing their brains to release the necessary fuel to maintain their focus and self-control in many other situations.

This new "unified" theory could solve many other mysteries about self-control and concentration that previously puzzled scientists. There had been signs, for instance, that people's mental focus increases if they are led to believe they are near the end of a task, but not if they feel they still have a long way to go—a finding that is hardly compatible with the original iteration of ego depletion theory, but which makes sense when you consider the brain's need to allocate resources. This theory might also explain why people generally perform better on mentally taxing tasks if they are paid to succeed. Given the immediate prospect of a reward, the brain—even one that subscribes to the limited view—is more willing to risk exhaustion by devoting more resources to neural activity.[27]

The brain's internal bookkeeping can also help us understand why the mere taste of sugary drinks immediately improves performance, even before the glucose can reach our neurons. Some studies have even found that rinsing the mouth with sugar water and then spitting it out can improve performance. Once the glucose receptors in the mouth have signaled that more fuel is coming, the accountant knows that the brain can afford to spend its existing energy more generously. Consistent with this idea, Job has shown that ingesting sweet, sugary drinks tends to offer more relief to people with the limited mindset—who are more likely to feel that they are on the verge of exhaustion, and whose brains are parceling out their existing reserves conservatively—than to people with the non-limited mindset.[28]

Many other substances used to aid concentration may work through belief rather than any direct effect of the chemicals themselves. When studied in controlled trials that properly account for the influence of expectation, the brain-boosting effects of caffeine were found to arise primarily from our beliefs about its benefits. Indeed, one study found that the mere smell of coffee is enough to lead to immediate improvements in performance, thanks to its associations with mental acuity. Even so-called smart drugs such as amphetamine salts—used by ambitious students and workers to enhance their concentration and focus—may work through altered expectations of our own capability, independently of any biochemical effects.[29]

## A WILL OF STEEL

Just how much willpower can any person use before they run out of gas? The astonishingly prolific American writer Danielle Steel may offer some surprising clues. One hundred and seventy-nine books into her career, Steel revealed that she had achieved her success by working twenty-hour days, beginning at eight-thirty in the morning, while resisting almost all distractions. If she faces a creative challenge, she relies on her boundless energy to keep on slogging through it. "The more you shy away from the material, the worse it gets. You're better off pushing through," she advised. Steel said that she fails to understand how she could become exhausted by her work, an attitude that certainly sounds very close to the non-limited beliefs about the mind that Job has studied. She even has a sign in her office that reads: "There are no miracles. There is only discipline."[30]

The interview, published in 2019, soon went viral, with many newspapers and other outlets describing Steel's powers of focus and self-control as "superhuman." As the *Guardian* pointed out at the time: "It's one thing to be driven. It's quite another to have to summon up and sustain the willpower to do the driving."[31] This response is not surprising, considering that most people believe even a short period of uninterrupted mental exertion is tiring.

Job's research suggests that many more of us could attain higher levels of productivity if we adopt the right mindset—but Steel's work ethic might ring some warning bells, too. One potential criticism of Job's research is that "non-limited" beliefs about the brain could lead us to push ourselves too hard, without permitting any pleasure in life. Fortunately, extreme workaholism does not seem to be a common problem for people with the non-limited beliefs, at least according to Job's studies. Indeed, the research shows that they tend to be happier and healthier than those who expect mental effort to be depleting, which is hardly what you'd expect of the typical workaholic. One reason is that they use their mental resources to plan their work effectively, without wasting time on distractions. "They are more efficient at reaching their goals, which is a strong predictor of well-being," Job told me. And once they have left work, they will feel more energized to take care of the rest of their life.

People with the limited beliefs, in contrast, tend to feel so depleted that they fail to organize their work and will feel more burdened as a result; they will then return home so exhausted that they find little energy to enjoy their time off either.[32] Surprisingly, people with the limited beliefs even suffer from worse sleep, since they don't have—or don't think they have—the self-control to get an early night when they need it and instead put off bedtime, a phenomenon called "sleep procrastination," which will compound their exhaustion.[33] The truth is that a boost to your willpower should help you to achieve whatever work-life balance is best for you, whether that's an all-consuming schedule like Steel's or a few hours of concentrated work followed by plenty of time for play. Your mental resources will be there for you to use whenever you need them.

If you have a limited mindset and want to change your own expectations, simply learning about your vast mental reserves can have immediate benefits for your focus and self-control. In one experiment, participants read an article about the "biology of unlimited willpower," which described the (typically) abundant reserves of glucose in our brain and the body's capacity to release more when we need it. The message became a self-fulfilling prophecy: their concentration and focus on cognitive tests actually *increased* with greater workload once they had been told about the brain's available resources.[34] Once again, merely taking a few minutes to reappraise your expectations can change something that was thought to be physiologically intractable.

As you process these new ideas, you might benefit from thinking a little about the times you yourself have felt energized after a demanding task in the past. Even if you have the limited beliefs about the brain, there will have been times when you felt you were "in the zone" on a complex task, so fully absorbed by what you were doing that you didn't feel time passing. You might have been engrossed in a novel or playing a complex computer game late into the night, for example. Those are both examples of your focus increasing with effort—even though you may not have recognized that at the time, since you were enjoying yourself so much. Or consider a moment when you found that exercising self-control left you feeling stronger. Remembering that kind of event can open your mind to the idea that your mental reserves of focus and self-control are much deeper than you assumed.

Once you have started to recognize this, you can test your limits with small challenges. They should reflect a realistic goal that you already feel highly motivated to achieve.[35] (In the classic ego depletion tests, people tend to find voluntary activities, picked with a sense of autonomy, to be less fatiguing than those imposed by others.) It could be as straightforward as avoiding the temptations of social media for a whole day, to test whether you can work more productively than you had previously imagined; if you tend to find your evenings are wasted with unfulfilling activities, you might try practicing a hobby rather than watching TV for one evening—to see if this leaves you more energized. Given Job and Savani's research in India, you might even try the traditional yogic exercise of trataka, in which you focus on a single point for a few moments to "cleanse" the mind and sharpen your concentration.

Whatever you do, don't start out by seeking to emulate Danielle Steel's prolonged displays of willpower: if you attempt too much too quickly, you may fail, and that will only cement the belief that your mental reserves are limited and easily depleted.[36] As we also saw with research on stress (Chapter 7), you should edge out of your comfort zone and then question how you felt throughout. With time, you may find that you are more capable of exercising self-control and concentration when you need it.

Parents and teachers may want to take particular note of these findings. In education, self-control and concentration are often as important for a child's academic success as natural intelligence, and Job's research suggests new ways that these qualities may be cultivated at a young age. The psychologist Kyla Haimovitz and colleagues recently visited a nursery school in the San Francisco Bay Area and read the four- and five-year-old children a short story about a little girl who had to exercise patience and determination—while waiting to open a present, buying an ice cream, or solving a difficult puzzle at school. At each challenge, the character felt "stronger and stronger" the longer she waited, or the more she persevered—a message that was designed to prime the children with the non-limited view of willpower.

After hearing the story, the children were given a classic test of self-regulation: the option of either having a small treat as soon as they felt like it or waiting for a full thirteen minutes to have a bigger treat instead (which is pretty much the hardest test of self-control you can give a preschooler). Overall, 74 percent of the children who had heard the inspiring story managed to resist temptation, compared to just 45 percent of a control group who heard a different tale.[37] Clearly, a single storybook is not going to change a life, but with regular exposure to similar messages, children should be better equipped to exercise their willpower in all kinds of tasks, providing them with a built-in resilience during times of pressure and exertion in later life.

In the next chapter we will examine more strategies through which teachers and business leaders can use expectation effects to enhance the potential of whole groups of students and employees. But to end our exploration of willpower, let's look at a final means of boosting focus and self-control—through prayer or ritual.

## SECULAR SUPERSTITION

Look into the biography of any elite performer in sports or entertainment and there is a good chance they'll have some kind of superstition or ritual. Most basketball players, for example, develop a routine set of movements that they perform—such as dribbling, pausing, bouncing the ball, and spinning it a certain number of times, or even kissing it—before taking their foul shot. Serena Williams listens to the same song ("Flashdance . . . What a Feeling" by Irene Cara) before she goes on court, and she bounces the ball exactly five times before her first serve; Rafael Nadal always takes a cold shower before each match and performs a characteristic series of gestures as he waits for his opponent.

In the arts, Beyoncé prays and performs a fixed set of stretching exercises before a performance, while the ballerina Suzanne Farrell—considered one of the greatest dancers in the United States—always pinned a small toy mouse inside her leotard. Superstitions and rituals are also prevalent among writers—Dr. Seuss would wear a lucky hat when he felt that he was blocked—and composers: Beethoven relied on coffee to fuel his creativity, and he religiously counted out exactly sixty beans for each cup.[38]

Before I learned about the power of expectation effects, I would have believed these superstitions were a kind of emotional crutch without any direct benefits to performance. And I would have been mistaken. One study of free throws in basketball found that players were around 12.4 percent more accurate when they followed their personal routines before the shot than when they deviated from the sequence.

Overall, the total success rate was 83.8 percent with the exact routine, compared to 71.4 percent without.[39] Superstitions and rituals can also boost perseverance and performance across a whole range of cognitive tasks, and the advantages are often considerable. In a test of verbal dexterity, for instance, the presence of a lucky charm brought a 50 percent improvement. Simply hearing a superstitious phrase—such as "break a leg"—can bring about a small advantage, according to one study.[40]

Why would rituals enhance people's skills—across so many different domains—in this way? One obvious explanation is that superstitious beliefs and rituals help to quell anxiety and create the sense that you have more control over the situation. That is almost certainly a factor. Equally important, however, they may lead us to have more faith in our mental reserves and our capacity to maintain concentration and self-discipline. As a result, we can persevere even when others would begin to feel exhausted and start to give up, and with the increased mental focus, we can avoid any distractions that might limit our performance.

One study that directly tested the effects of superstitious beliefs on willpower found that people who spend time in spiritual contemplation sustain greater focus on tests of concentration than those who do not.[41] If you are beginning to feel your discipline weakening, the belief in a helping hand from some supernatural power can replenish your reserves.

Given these results, some researchers have speculated that the boost for willpower could have been a primary reason that many cultures evolved religious beliefs and rituals in the first place.[42] In our evolutionary past, the increased self-control might have helped them to rein in their worst impulses (such as aggression or stealing from neighbors) and forgo immediate pleasure (such as gorging on restricted food supplies) for the future good of the group.

Fortunately for atheists, you can still benefit from certain rituals without any need for a higher power. As we saw with open-label placebo studies in Chapter 2, placebo treatments can be effective even when participants are fully aware that they are taking the sham drug—apparently because the routine itself appears to trigger the expectation that they will get better. Superstitions and rituals seem to be no different, and there is strong evidence that they can provide a boost even if people are fully aware that there is no rational reason why they should work.

In one beautifully bizarre experiment, Alison Wood Brooks at Harvard University and colleagues invited participants to perform a karaoke rendition of "Don't Stop Believin'" by Journey. To ensure that they made their best possible effort, the subjects were told they would be judged for the accuracy of their performance, as rated by the karaoke software, and they would be given a bonus of up to $5 for a pitch-perfect rendition of the song.

Before the performance began, around half of the participants were given the following instructions:

> Please do the following ritual: Draw a picture of how you are feeling right now. Sprinkle salt on your drawing. Count to five out loud. Crinkle up your paper. Throw your paper in the trash.

The mere act of performing the ritual—which could have offered no direct benefit to their singing—boosted the participants' singing scores by 13 points out of 100, compared to control participants who had waited quietly until they were asked to perform. Follow-up experiments revealed similar improvements on a difficult math test. And the exact framing of the preparatory routines mattered. If it was called a "ritual," the participants saw the benefits, but not if they were asked to perform "a few random behaviors." The connotations of the word clearly mattered in boosting their ability to remain focused under pressure, in just the same way that the word "placebo" brings its own medical benefits.[43]

The adoption of secular rituals can also improve our resolve in some classic tests of willpower that often leave us feeling depleted, including our capacity to resist the temptation of a tasty treat. Participants in one experiment were asked to perform some ritual gestures (sitting upright, closing their eyes, bowing their head, and counting to ten) before they ate, while a control group performed a set of random movements. They were then given a choice between a Snickers bar and a lower-calorie fruit-and-cereal bar.

On a subsequent questionnaire the participants who had performed the ritual were more likely to report greater feelings of discipline, giving higher ratings to statements such as "I felt mentally strong when making this decision," and "I felt sharp and focused when making this decision." And this was reflected in the food they chose to eat. Around 64 percent of those in the ritual condition chose the healthier option over the chocolate bar, compared with 48 percent performing random movements without the sense of ritual.[44]

Given these results, we might all consider adopting a few rituals that help to engender a sense of control and focus. The aim should be to choose something personally meaningful and straightforward: like the dieters performing the ritual movements, you want something that can easily evoke feelings of inner strength. (If it is overly complex, the routine may itself be difficult to maintain in the hour of need—creating a burden that could potentially add to your anxiety and lower your performance.)

It could be as simple as performing exactly the same sequence of stretches each morning before work, or having a particular voice warm-up that you repeat before an important presentation, or saying a special mantra before your discipline is about to be tested. I personally try to build a ritual around my morning coffee, before I start writing—counting out the number of beans like Beethoven to imbue the drink with a sense of significance and to prime my mind for focused concentration. If you have some favorite clothes or perfume, turn that into a lucky charm that you use when you know you are going to have to act under pressure.

Whether you are a professional athlete, singer, or public speaker, or you just want to have the self-control to avoid procrastination and stop wasting time, the only thing holding you back may be your own expectations of your willpower. And a little manufactured "luck" and sense of control could be all you need to set you on the path to success.

### HOW TO THINK ABOUT . . . WILLPOWER

- A sense of autonomy—the feeling that you have control over your activities—can reduce the sense of ego depletion, even if you have the limited mindset. Where possible, make an effort to establish your own routines rather than following others' orders, and regularly remind yourself of their purpose and meaning to you personally.
- Try to recognize the instances when you personally find mental effort to be invigorating. What difficult tasks do you enjoy *because* of their difficulty? Reminding yourself of these activities will help you to build up your beliefs in your potential.
- When you find an activity to be draining, consider whether it is objectively more difficult than the things you find energizing, or whether it is just a preconception. Do other people find it energizing, for instance, and is it objectively more difficult than the other activities that do not feel so depleting? By questioning these assumptions, you may begin to see that you are capable of much more than you thought.
- Establish your own rituals and secular superstitions that will help you to establish a feeling of control in times of high pressure. It could be a "lucky" charm that has positive associations or a set of reassuring gestures—anything that feels personally significant and brings the promise of success.

# 9

# UNTAPPED GENIUS

## How to boost your own—and others'—intelligence, creativity, and memory

Take a moment to consider the people around you—your boss, your colleagues, your partner, and your friends. Do you feel smart when you're around them? Or do they leave you feeling slow-witted and unoriginal, so that you are always struggling to play catch-up? How about the people from your past—like your schoolteachers or your parents? Did they see your potential? Or did they underestimate you?

In Chapter 4 we saw how we could catch psychogenic illnesses from others through contagious nocebo effects. Now that we've examined the ways that our beliefs can influence our resilience and willpower, it's time to examine how the beliefs of the people around us can change our intellectual abilities. Whenever we interact with another person, they can transmit their opinions of us through subtle cues, and over time we can come to internalize those expectations as if they were true—leading to profound changes in our performance. If you have ever found that some people bring out the best or worst in you, this is why.

The first clues to the existence of this phenomenon came from a seminal experiment at Spruce Elementary School in South San Francisco.[1] It was the spring term of 1964, and the staff were busy with their own packed schedules, when their principal Lenore Jacobson asked them to handle one more demand. A psychologist called Robert Rosenthal, she said, wanted to identify the children who were on the cusp of a sudden "spurt," during which they would show accelerated growth compared with their peers. To do so, he had designed a cognitive test that could predict the child's trajectory, and he hoped to try it out at the school. Each child completed the exam, and after the summer term each teacher was given a short list naming a small number of children who were most likely to be "bloomers."

As you might expect, the supposed premise of the research was, in fact, a sham. The bloomers had been picked at random to see whether the teachers' enhanced expectations would influence the children's progress over the following year. The test that the children had been given in spring 1964 served as a baseline to measure those intellectual gains.

And for some alleged "bloomers," the effects of the teachers' enhanced expectations seemed truly remarkable. There was Violet, a "small, wiry tomboy with little black eyes." She was the fifth of six children, the daughter of a butcher and a housewife. The staff across the school knew her for her defiance and her playground fights. Despite these behavioral issues, however, her intelligence developed tremendously over Grade 1, with a second test revealing that she had gained 37 IQ points. That's an enormous increase in intellect that would have seemed impossible with intense individual tutoring, let alone a standard elementary education.

Then there was Mario—the son of a factory worker and a typist, who was just starting Grade 2. He was already known to be a bright little boy, even if his reading aloud was sometimes faltering, and he still wrote letters as their mirror image. Eight months after the first test, however, his intelligence had grown by the equivalent of 69 IQ points.[2]

Not all the children showed such remarkable progress. Overall, however, the bloomers' intellectual gains were about twice those of the other children in their year, outstripping their classmates by 15.4 IQ points in the first grade and 9.5 IQ points in the second grade.[3]

Importantly, the teachers weren't simply paying more overt attention and giving more care to these children. If anything, they were spending *less* time with them. Instead, the teachers seem to have subtly communicated their beliefs through daily interactions, which, in turn, led the children to take a more positive view of their own abilities—beliefs that allowed their young minds to flourish.

Rosenthal and Jacobson's results were initially considered to be controversial. With our newfound understanding of the expectation effect, however, the bloomers' progress makes perfect sense. Our intellectual performance can be influenced by our beliefs, and we often absorb assumptions from the people around us. Too often those expectations act like brakes that slow our progress, but once you release them, it suddenly becomes much easier to reach your potential. As we shall see, the implications of this research are political—since there is strong evidence that expectation effects can increase social equality. Fortunately, some cutting-edge techniques allow us to break free from the limits imposed on us by others, so that we can create our own self-fulfilling prophecies.

## INSTANT BRAIN BOOST

The very idea that we could "think ourselves smart" is itself rather shocking. For much of psychology's history, our intelligence has been considered a prime example of the nature versus nurture debate. Our genes are supposed to be the biggest factor determining brainpower, followed by factors like diet and the home environment. The effect of expectations should be marginal.

Some contrary evidence appeared to come from the science of brain training. If you were interested in computer games in the late 2000s, you may remember seeing a raft of apps that claimed to increase your intelligence. The most famous games were Dr. Kawashima's Brain Training, which was available for the Nintendo DS and was publicized with widespread advertisements by the actress Nicole Kidman; and Lumosity, a website and app that since its inception has had more than a hundred million users.

Much like Roy Baumeister's theories of willpower, the argument was that your brain operates like a muscle—and the more you exercise it, the smarter you become. The apps typically included games that were designed to increase your working memory, spatial reasoning, cognitive flexibility, and mental arithmetic—skills that together are thought to determine your intelligence in a range of different tasks. Users often reported increased mental clarity and sharpened memories, and the academic literature appeared to prove the point, with users showing noticeable differences in IQ after a few weeks of regular training. Maybe nurture—even in adulthood—could give nature a run for its money after all.

Unfortunately, many of these studies did not involve an "active" control—that is, a suitable comparison that might lead people to believe they are making some kind of useful effort.[4] And for those studies that did involve a control, the activity was often uninspiring—such as watching an educational DVD—which may not have evoked the same feeling of mental engagement as playing an interactive game.[5] We've all attended boring classes, after all, without noticing a sudden brain boost afterward. As a result, the expectations of improvement in each condition were probably very different. Even more problematically, the recruitment of the participants—most often university students—had often primed them to expect some kind of mental improvement, by overtly advertising the activity as a "brain training experiment." As a result, the students arrived at the labs with some very strong assumptions about what they would experience.

To find out whether an expectation effect might have skewed the earlier results, Cyrus Foroughi and colleagues at George Mason University in Fairfax, Virginia, set about recruiting students using two separate flyers, which they placed throughout the university campus.

The first very explicitly set up expectations of a large brain boost:

BRAIN TRAINING AND COGNITIVE ENHANCEMENT
Numerous studies have shown that working memory training can increase fluid intelligence.
Participate in a study today!

The second instead focused on the incentive of gaining college credits:

EMAIL TODAY AND PARTICIPATE IN A STUDY
Need credits? Sign up for a study today and earn up to 5 credits.
Participate in a study today!

Each leaflet offered a different email address to register interest in the study, allowing Foroughi and his colleagues to determine which message each participant had seen before the test. Once they arrived at the lab, the participants took two separate intelligence tests—providing a baseline score before they engaged in an hour's brain training. They were then given a night's sleep before taking two further intelligence tests on the following day.

As Foroughi himself points out, it would be highly unlikely that a single hour's training could have a meaningful effect on intelligence. (A whole year of further education, after all, is only thought to add at most around 5 points to someone's IQ.) Yet that is exactly what his team found in the high-expectation group, with those participants showing a huge gain of 5–10 IQ points on the two tests, while the control group showed practically no improvement. Through the power of expectation, the first group had experienced an instant brain boost.

For further evidence, Foroughi looked back through previous brain-training experiments to check whether or not they had offered overt advertising of the kind he'd used in his study, and to compare the size of the effects to those experiments in which the recruitment effort had not explained the proposed benefits. Sure enough, he found that the cognitive gains were much greater for the experiments that had (unwittingly) raised the participants' expectations.[6]

Foroughi's study was taken by some to be evidence that brain training simply doesn't work. That conclusion was an oversimplification. In reality the study confirmed that heading to the "mind gym" and engaging in tough mental activities really can strengthen your brain, at least in the short term—but part of that success comes from the hype. In her ads for Dr. Kawashima's Brain Training, Nicole Kidman actually was making all the users a little bit smarter.[7]

Similar expectation effects have now been documented in tests of noninvasive brain stimulation. You can buy devices that apply small electrical currents to the scalp, which are supposed to change the activity of neurons underneath—and they are sometimes touted as offering an instant brain boost. There is still an academic debate over whether the technology is as powerful as some claim, but at least some of the effects seem to arise from people's assumptions about the interventions and their own capacity to improve.[8]

IQ is only one way of measuring our intellectual potential, of course. It's perhaps best seen as an underlying brainpower that determines how quickly we process new information. But we now know that many other intellectual capacities are also susceptible to expectation effects.

Consider creativity and the capacity to come up with original solutions to problems. Read any business magazine or website and you'll soon come across an article offering tips on the best ways to improve the originality of your ideas—from taking a shot of vodka[9] to lying down.[10] Like the brain-training studies, however, the experiments often fail to consider people's beliefs on entering the lab. Most of us know, after all, that writers like Ernest Hemingway sometimes had their greatest flashes of inspiration while under the influence, meaning that when we're given a drink and asked to take a test of creative thinking, we're already primed to do better; similarly, we may have heard that many writers, such as Truman Capote, Vladimir Nabokov, or (more recently) Phoebe Waller-Bridge, have preferred to work in bed. Those assumptions are highly likely to have influenced the results of studies, and, without adequate controls in place, we can't know if it's the vodka (or supine posture) or our expectations about what vodka (or supine posture) will do for us that makes all the difference.[11]

To see if enhanced beliefs can make us more creative, a team from the Weizmann Institute of Science in Rehovot, Israel, primed people with the expectation that a sniff of cinnamon could help them to come up with more original ideas. As expected, the participants scored much higher on a standard measure of creativity, in which they had to suggest new and innovative uses for common household objects—such as a shoe, nail, or button—when they were exposed to the smell. Participants who had only been given the sniff test, without any expectation that the perfume could lubricate their thinking, did not see these benefits.[12]

How about memory? Your expectations can't create knowledge where there is none already: contrary to some more far-fetched claims of the mind-body-spirit literature, you can't just tell yourself that you are fluent in French and speak like Audrey Tautou. But many of us have absorbed more information than we realize, and a recent study of students' general knowledge shows that an expectation effect can make it easier, or harder, to retrieve those facts.

The researchers led the participants to believe that they were involved in a test of subliminal messaging, and that the answer to each question would be flashed momentarily across the screen. If they were asked, "Who painted *Guernica*? Picasso, Dalí, Miró, or El Greco," for example, then Picasso should appear just beforehand. The participants were told that the writing would disappear before they were consciously aware of its presence, but their subconscious mind would be able to pick it up. "Follow your intuitions," the researchers told them.

"On some level you already know the answer." In reality, of course, there were no subliminal clues—but the belief that they had received a helping hand meant that the participants were much more likely to pick the correct answer each time.[13]

How could this be possible? Neuroscientists studying consciousness often talk about our "mental workspace"—which you could picture as a kind of whiteboard that allows you to juggle a limited amount of information at any one time. If you consider yourself to be a relatively slow-witted and unoriginal thinker, then your anxieties about your abilities will be cluttering that workspace whenever you are required to practice them. If you already have more faith in your intellectual potential, however, then that workspace will be clearer for you to hold more information—and your thinking is going to be more focused and less inhibited, allowing you to devote more attention to the specific task at hand. That belief in your own abilities will also mean that you are more likely to persist even if the best solution doesn't come immediately to mind.

Expectation effects are especially important when we confront new difficulties that challenge our existing abilities. Abundant research shows that moderate frustration—which often accompanies fresh challenges—is actually a sign of learning; if you find something difficult to comprehend or execute, you are more likely to make lasting improvements in your skills, or to remember the fact in the future, than if it is immediately easy. (It is for this reason that neuroscientists tout the benefits of "desirable difficulties" for learning.) Unfortunately, most of us struggle to realize this, and we instead start to develop the fear that we'll never improve, which becomes self-defeating. Some powerful experiments have found that reminding yourself of the benefits of frustration can remedy those problems, reducing the feelings of helplessness and freeing up mental resources like working memory so that you will begin to perform better over time.[14] Once again, it's a self-fulfilling prophecy: if you expect frustration to help you learn, it will;

if you think frustration is a sign you're out of your depth and always will be, it is.

You must beware the trap of overconfidence, of course—assuming that you are brilliant at everything, without any foundation for that self-belief, could also set you up for failure and embarrassment. The aim is to be realistic, preferably testing your abilities in small incremental steps. Don't begin with inflated expectations of huge leaps in your abilities, but simply question your assumptions and keep an open mind. Even if you do not usually go so far as to wallow in feelings of inadequacy, you might still assume that certain skills "are not my thing"—but once you probe the origins of those beliefs, you may find that you are actually a lot more capable than you'd once imagined, and you may be able to release hidden potential that will improve your performance in the future.

## THE POWER OF PYGMALION

If our own beliefs can constrain or release our intellectual potential, how about the beliefs of others about us?

After their study at Spruce Elementary, Rosenthal and Jacobson detailed their findings in a book called *Pygmalion in the Classroom*. The title was a nod to a classical myth told in Ovid's *Metamorphoses*, in which a sculptor falls in love with a statue he had carved, leading the gods to grant his wishes and bring her to life, and to George Bernard Shaw's play *Pygmalion*, in which a flower girl learns to act like a member of the aristocracy thanks to her passionate teacher. (*Pygmalion* is also the basis for the musical *My Fair Lady*, starring Audrey Hepburn, which was—perhaps coincidentally—released the year the study took place.)

Back in the 1960s the idea that beliefs could change outcomes really did seem to be in the realm of fiction and mythology. Remember that in this period expectation effects were still largely constrained to medicine, and even there they were seen as a distraction from the "real" physiological actions of drugs. It's little wonder, then, that contemporary psychologists were initially skeptical of the huge gains at Spruce Elementary. And there were some legitimate reasons to question the results—including the relatively small sample of students, which may have inflated the apparent size of the effect. Over the subsequent decades, however, many more studies have confirmed that a teacher's expectations can have either positive or negative influence on a child's academic performance.[15] If—for whatever reason—a teacher decides that a pupil is less able, they will, quite unwittingly, place brakes on that pupil's development, irrespective of the child's actual ability. Indeed, and unfortunately, the research suggests that these losses in performance may be more drastic than the brain gains that come from a teacher's positive views.[16] "I don't think there's now any doubt that teacher expectations make a difference," Christine Rubie-Davies, a professor in the faculty of education at the University of Auckland who has extensively investigated the Pygmalion effect, told me.

In the last decade there has been an explosion of new interest in the Pygmalion effect, with evidence that the consequences of a teacher's expectations can be surprisingly long-lasting. In the early 2010s Nicole Sorhagen, then a PhD student at Temple University in Philadelphia, analyzed the results of a survey tracking the progress of a thousand children in ten cities across the United States. In the children's first grade, their teachers had been asked to rate the children's various academic skills—subjective opinions that Sorhagen was then able to compare with the children's actual performance on standardized tests in the same year. If there was a discrepancy between the two, it would suggest that the teachers had unfairly high or low expectations of the child. She found that those early judgments could predict the children's math grades, reading comprehension, and vocabulary at the age of fifteen. The advantage of a teacher's high estimation, or the disadvantage of an unreasonably low appraisal, remained with the child through to high school.[17]

Perhaps unsurprisingly, the idea has also caught the attention of organizational psychologists looking for ways to increase productivity, and it is now clear that others' expectations can be a powerful force in the workplace, determining the performance of everyone from Dutch police officers to the administrators at a large US bank. In each case, a leader's expectations boosted or limited their employees' performance in just the same way that the teachers at Spruce Elementary had released their pupils' intellectual potential.

The most impressive study examined Israeli soldiers undergoing a fifteen-week combat command course, during which they were tested on various tactical and practical skills.[18] Priming a leader to have positive expectations of a particular trainee was found to improve the trainee's mean score by three standard deviations. This means that the most average soldier under normal circumstances would rise to the top 0.1 percent of recruits if the trainer believed them to have high potential.

Such huge gains are exceptional—there seems to have been something about the Israel Defense Forces that was especially conducive to this expectation effect, and it should not be expected elsewhere—but the average effect size across professions is still very large compared to most psychological interventions, leading the average person to rise about 16 percentiles within their group if their leader has a positive view of them.[19]

Exactly how the expectations are transmitted from the teacher or leader to their student or employee will depend on the people involved and the specific situation. The most obvious means would be overt praise or criticism; we all know that encouragement can be helpful and criticism hurtful. But someone's expectations are also evident in the goals they set, which can then affect performance. If a teacher continually chooses more ambitious tasks for their favorites, that provides further opportunities for learning, while the rest of the group misses out on those opportunities.

Other signals may be subtler. Imagine you are asked a question and make an error while answering. If someone has high expectations of your abilities, they might rephrase their question or talk you through the problem. Someone with lower expectations, however, may simply move on, which subtly hints that they don't think you are going to learn from the mistake.[20]

Perhaps most important are the nonverbal cues. People are less likely to smile, and they make less eye contact if they have lower expectations of you, for example—small differences in interaction that are nevertheless easily perceived by children and adults. Even silence can be important. If someone leaves a short pause after you have given a quick response to a question, it can give you a further chance to expand on your ideas and refine your thinking. Psychologists describe these subtle signals as "leakage," since people can accidentally communicate their expectations even when they are trying to conceal their feelings.[21]

However the expectations are communicated, the research shows that they are soon internalized by the people on the receiving end, reducing or raising their motivation and self-belief. Often we may not even be consciously aware of the cues that cause us to feel this way—but those feelings will affect our performance.

We might hope that with time and hard work we could eventually prove ourselves, and that a teacher's or leader's beliefs about our abilities would update accordingly. Unfortunately, exceeding someone's expectations can have counterintuitive consequences. While the teachers at Spruce Elementary School embraced signs of precocity in the bloomers, Rosenthal and Jacobson found that the teachers took a harsher view of other children who had not been labeled as bloomers yet nevertheless made unexpected progress. "The more they gained, the more unfavourably they were rated," the psychologist and the principal noted in their write-up for *Scientific American*.[22] Once the teachers had formed a negative belief, the child faced an uphill struggle to please them.

This may be another form of the human mind's "confirmation bias":

we are always looking for reasons to support our existing opinions, and when the contradictory evidence is quite literally staring us in the face, we will choose to dismiss it rather than updating our beliefs. Like a dramatist carefully crafting a narrative arc, we don't like the objects of our expectations to go off script.[23]

## THE BIAS BUBBLE

The consequences of the Pygmalion effect would not be so terrible if the majority of people were accurate and fair in their judgments of others. In most autobiographies, after all, you'll hear how a mentor saw the amazing potential in the young star and—through their relentless encouragement—helped them to recognize their talents. And in these few selected cases, the mentors were entirely justified.

For the polymath Maya Angelou it was Bertha Flowers whose continued encouragement inspired Angelou's love of literature and empowered her with feelings of self-worth. "I was respected not as Mrs. Henderson's grandchild or Bailey's sister but for just being Marguerite Johnson," she wrote.[24] (Marguerite Johnson was Angelou's name at that time.) For Oprah Winfrey it was Mrs. Duncan. "I always, because of you, felt I could take on the world," Winfrey told Duncan, face-to-face, on her chat show.[25] And for the physicist Stephen Hawking it was Dikran Tahta, who saw through the young student's poor handwriting and natural laziness and inspired his fascination with the universe. "Behind every exceptional person, there is an exceptional teacher," Hawking claimed.[26]

Unfortunately the psychological research reveals that most people's capacity to detect a latent talent is far from exceptional. We judge others and are judged in turn on small superficial differences, meaning that the advantages of someone's warm expectations are unfairly bestowed on some, while others are completely overlooked.

Just consider a bias known as the "halo effect," which causes us to assume that people with more symmetrical faces—essentially those who are more stereotypically attractive—are also more intelligent and competent. There is no logical reason for this; it's pure prejudice. In the words of one paper's authors, we are "blinded by beauty."[27]

Sadly, we are judged on our appearance from a young age—and when those expectations are continually transmitted to us through our parents, teachers, coaches, and managers, they can determine our actual performance on many tasks for which our looks should be completely irrelevant. The distorted perceptions eventually become our reality.[28]

As the research on the general Pygmalion effect has shown, the consequences can add up over time.[29] Various studies have confirmed that a child's looks can predict their teacher's expectations and, in turn, their academic achievement.[30] And if you do slightly better at school then you might also get better work—which can again be aided by people's judgments of competence based on your looks. You will then be more likely to get a promotion and to earn a higher salary—and the benefits can snowball over time. A study from the early 1990s, for instance, found that ten years after graduating, an attractive MBA student earned around ten thousand dollars more per year than the least attractive people in the class.[31]

Even something as supposedly inconsequential as the pitch of our voices could influence our long-term success through the cumulative advantages (or disadvantages) that arise from the halo effect. In general, people with deeper voices are considered to be more competent. When William Mayew at Duke University in Durham, North Carolina, analyzed voice recordings of 792 CEOs from some of the USA's largest companies in 2013, he found that those with the deepest voices tended to control the largest companies. All other things being equal, they earned $187,000 more per year.[32] The pitch of their voice also appeared to influence their tenure—how long they were allowed to stay in their position.

(James Skinner, the CEO of McDonald's, had one of the lowest voices in the sample. His average yearly earnings were $14.71 million.)

It is astonishing—and frightening—that such superficial differences can have such a powerful influence over others' perceptions of us, and, as a result, over the trajectory of our lives. And there are many other, far more important ways that expectations based on stereotypes could be shaping people's actual abilities.

The British neuroscientist Gina Rippon, for instance, argues that adults' expectations about gender roles start to shape their brains from the moment they are born. She describes babies as "social sponges" and argues that even subtle cues from parents, teachers, or friends could boost or diminish their burgeoning abilities in different fields by bolstering their confidence or creating a sense of anxiety.

A relative might show slight surprise when a girl plays with a LEGO set, for instance, subtly signaling that this is unexpected and—in the child's interpretation—undesirable, and so she may be less likely to play with the toy in the future. That lost play time may seem inconsequential, but playing with LEGO bricks would have helped to train her in spatial and nonverbal reasoning skills—meaning that when the girl grows up she will be at a slight disadvantage compared to boys of a similar age.

At school, meanwhile, an adult might plant the seed of doubt in a girl's head that makes her underperform in a test. Having had her fears confirmed, she may then perform even worse in the next test, making her less and less enthusiastic about the subject. Those initial expectations could thereby harm her immediate performance and disrupt her long-term learning until—for this child at least—the idea that "girls can't do math" has become a self-fulfilling prophecy.

Such prejudices could extend beyond education and into the workplace, making everything a little more challenging than it needs to be, and over time they could have contributed to the gender gap in science, technology, engineering, and math careers.

Some people still deny the importance of expectation and instead argue that gender differences are innate. These skeptics will point to brain scans that apparently show some kind of anatomical difference between the brains of boys and girls, or men and women. There are claims that males have larger brain regions associated with spatial reasoning or numeracy, for example. Far from demonstrating an inherited difference, however, the anatomical variation shown in these brain scans is a reflection of our culture's gender bias. It is natural that the brain responds to its environment and the skills that we have been encouraged to practice. If you are a child playing with LEGO bricks, you are actively changing your brain's wiring. As a result, these supposed differences are simply another illustration of the ways that our expectations—and those of the people around us—can have a real, physical effect on our biology.

Besides the apparent gender gap in aptitude for certain academic realms, expectation effects also exacerbate the consequences of economic inequality. There is now strong evidence that teachers consistently underestimate the abilities of poorer children. This is particularly unfortunate, since the research shows that children from working-class backgrounds could benefit the most from teachers' positive expectations, with the boosted brainpower helping to make up for the reduced resources at home.[33] If your circumstances are already set against you, you need every ounce of confidence you can receive.

The consequences of flawed expectations are a particularly serious problem for ethnic minorities (a fact that Jacobson and Rosenthal had indeed noted in their original writing on the subject). There is abundant evidence that many people who are not overtly racist hold implicit prejudices based on ethnicity, and that these biases can be communicated nonconsciously—with important consequences in academia and the workplace. Proving this point, one major survey in the United States tracked a diverse sample of more than 8,500 pupils from kindergarten to eighth grade, with a specific focus on their performance in math. The study, published in 2018, concluded that the effects of teacher expectations "are stronger for white girls, minority girls, and minority boys than they are for white boys."[34] Often these low expectations can permeate the whole culture of an organization. Staff at poorer, more ethnically diverse schools are more likely to think that these students as a group are "less teachable"—an attitude that will then prove itself to be true, thanks to the teachers' own behavior as much as the students' difficult circumstances.[35]

You'll sometimes hear arguments that minority or disadvantaged groups should strive to "lean in" and overcome these barriers, working hard to disprove the prevailing cultural biases. That's a big ask though. When people feel themselves to be at risk of conforming to negative expectations about their group, they experience a type of anxiety known as "stereotype threat," which damages their performance. Reflecting on the predicament, or even psyching yourself up to work harder because you've been told something like "you'll need to work twice as hard to get half as far," may only increase that stress.[36]

As children, we were thrilled by fairy tales in which a random child was blessed or cursed by a fairy godmother or a witch—but the alarming reality is that many people are subject to other's prophecies based on nothing more than their race, gender, or the appearance of their face, and even the subtlest biases have the power to change the trajectory of their whole life. If we continue to let low expectations go unchecked, there will be many people with hidden talents who never feel the kind of encouragement experienced by Hawking, Winfrey, or Angelou; who never have an exceptional teacher to bring out the exceptional individual. And the world will be poorer for it.

So what is to be done?

## REWRITING THE SCRIPT

When Rosenthal and Jacobson first published their study, some commentators assumed that they were trying to erase, or at least discount, the other potential sources of inequality. "If thousands upon thousands of children are not learning to read, write, speak and compute, it is not because of overcrowded classrooms, the effects of poverty and social conditions, poorly developed educational programs and materials and inadequately trained teachers," one *New York Times* columnist opined sarcastically. "No, the children are not learning because the teachers don't expect them to learn."[37]

It was an unfair exaggeration of Jacobson and Rosenthal's views, but the importance of structural factors is worth taking seriously. In addition to the effects on them of other people's implicit expectations, many people still have to contend with overt sexism, racism, and classism, not to mention the institutional barriers that maintain inequality in the United States, the UK, and many other countries. Addressing expectation effects will not magically resolve these issues—any more than we would expect a placebo to miraculously cure a terminal illness.

Recognizing this does not mean that we should abandon the possibility and importance of psychological change though. Five decades on from that memorable experiment at Spruce Elementary, a growing body of research has shown that teachers and leaders can change the ways that they communicate their expectations of others. Although they may not be a panacea, these interventions will be an important first step to maximizing everyone's potential.

Christine Rubie-Davies, at the University of Auckland in New Zealand, led one of the most robust attempts to rewrite teachers' scripts without deception. Working with Rosenthal, she invited a group of ninety elementary and middle-school teachers to take part in a randomized, controlled trial. Half the teachers were given regular professional development sessions examining general ways to improve student engagement and achievement, while the rest attended four workshops that looked specifically at the importance of self-fulfilling prophecies. (Both the "Pygmalion" workshops and the standard professional development sessions were roughly equivalent in terms of the time and effort required.)

At the Pygmalion workshops, Rubie-Davies first educated the teachers about the power of expectation effects and the ways they could affect academic performance, as well as some strategies to increase *all* the students' expectations of themselves. These techniques included things like working with each student to set clear goals, establishing measures to make sure that regular feedback is given to everyone (and not just the favorites), and finding ways to encourage student autonomy—making it clear to students that it is often within their own power to solve their own problems.

The teachers were also asked to film themselves in the classroom—videos that were then analyzed in the subsequent workshops. The videos were the crux of the trial's success, since they allowed the teachers to identify the many ways that their low expectations could still be "leaking" through unconscious gestures, such as their body language and tone of voice. As the previous work on the Pygmalion effect had shown, the teachers were often completely unaware of their biases. "They suddenly realized that they only asked the boys the math questions, or they mostly interacted with the white kids," Rubie-Davies told me. "It became a really powerful learning experience."

The results were exactly as hoped—leading to a 28 percent improvement in the math performance of students whose teachers had gone to workshops, compared to the students whose teachers had undergone standard teacher training without any specific focus on the Pygmalion effect.[38] Clearly it *is* possible to change the way someone communicates their expectations of others, and to teach them how to empower others with greater self-belief—and that can have a significant impact on their (or our) lives.[39]

Ideally, these kinds of interventions would be common in every educational institution and workplace—and perhaps they will be when the concept of the expectation effect is widely known. In the meantime, it is at least possible to buffer ourselves against others' expectations.

Since many of the negative effects are caused by performance anxiety—and the imposed expectation that we are somehow not up to the task—we can use some of the stress reappraisal techniques covered in Chapter 7 to rethink the challenges we face. While these methods were designed for all kinds of anxieties, the available evidence suggests that they are especially effective at countering negative stereotypes. When girls were taught the potential benefits of anxious feelings—to energize the brain and increase performance—they tended to achieve better grades on math tests, for example. Importantly, the intervention brought the greatest effects when the girls had been explicitly reminded of the prevailing expectation that they would perform worse. The stress reappraisal had helped them to neutralize the stereotype threat.[40]

Alternatively, you might engage in a process called "self-affirmation." The name may sound off-putting, like an exercise from a New Age manual, but don't let that deter you. Self-affirmation, as defined by experimental psychologists, is not an act of wishful thinking, but a useful method to neutralize some of the more unreasonable doubts that you may be harboring.[41]

Rather than paying attention to the particular task in question—which may only trigger negative rumination about the expected difficulties—the aim is to focus on our general abilities and values quite unconnected to the problem at hand. Acknowledging those other personal qualities reinforces our belief in our own resources, while also reminding us that our self-worth needn't hinge on the particular challenge we're facing. Less anxiety, in turn, frees up the mental workspace from the negative thoughts that would be impeding our success, improving our memory and concentration—and helps to steel our resolve to continue through difficult challenges. You could think of self-affirmation as strengthening the foundations of our self-worth, so that our ideas about ourselves are no longer so easily swayed by others' opinions.

Try it for yourself now, and you'll see how easy it is. First, list ten characteristics—such as sense of humor, creativity, independence, social skills, or athletic ability—that personally matter to you. Now take the most important one and briefly describe why it matters, including a description of a time when it proved to be especially significant in your life.

This is the kind of short exercise you could perform anytime, anywhere—but its simplicity belies its power. In one of the first and most striking demonstrations, scientists at the Universities of Alberta and Arizona asked participants to perform a test of spatial awareness in which they had to match rotated shapes. As noted before, spatial skills are often assumed to be a weak spot for women—just think of all the sexist jokes about women reading maps—and thanks to the expectation effect, this often does become a self-fulfilling prophecy.

Before the test, half the participants were asked to complete a brief exercise in self-affirmation, while the control group was instructed to write about another person's characteristics. To check whether self-affirmation could help even when the negative beliefs were highly salient, the experimenters deliberately reminded all of the participants of the sexist stereotype, saying, "One thing we'll look at is how men and women differ in their performance on the test, and how true the stereotype is, or the generally held belief is, that women have more trouble with spatial rotation tasks."

The effects of the self-affirmation were remarkable: the participants who completed the intervention almost completely closed the gender gap in performance.[42]

The researchers found very similar patterns for math scores. Men, in general, do not need their self-belief to be bolstered (and so only

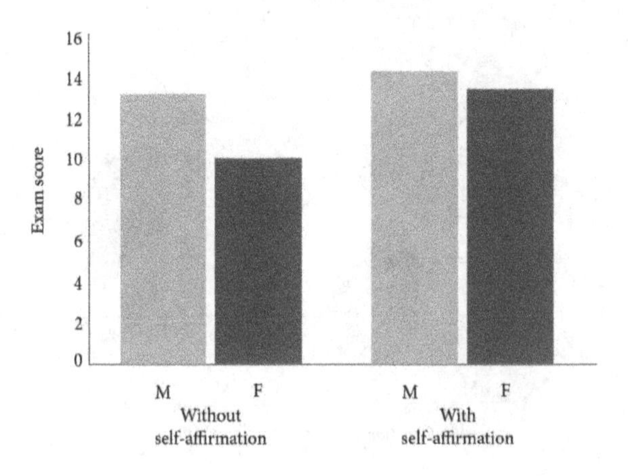


showed modest gains)—but women show a marked improvement after practicing self-affirmation.

If you're still not convinced, take a look at the graph overleaf, which shows the results of students taking a college-level introductory physics exam—another arena in which women are typically expected to underperform. On the left is the control condition, those who had not taken part in an intervention. On the right are the results of participants who had completed a couple of self-affirmation exercises at the beginning of the term and shortly before the midterm exam. As you can see, the gender gap had closed from around 10 percent to just a couple of points. Confirming the protective effects of the pre-exam exercise, the researchers found that the benefits of self-affirmation were greatest for the women who had previously bought into the sexist stereotype; in some way, self-affirmation was acting as an antidote to the negative beliefs transmitted by society.[43]

Besides reducing the gender gap, self-affirmation can also remedy the negative expectations that are often associated with poorer economic

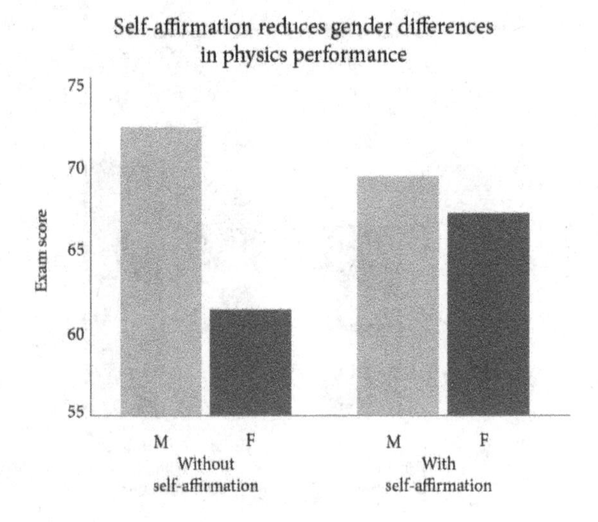


circumstances. Researchers in the UK asked eleven- to fourteen-year-olds to write a self-affirming essay in an English class at the start of the academic year. Comparing the academic results of children receiving "free school meals" (a form of government support to poorer families) with those of children from more affluent homes, the researchers found that the exercise shrank the class differences by 62 percent.[44]

The most breathtaking findings concern the academic achievement of Black students in the US. Like the UK students, Black students in this study were asked to practice self-affirmation at a special session at the start of seventh grade, with additional "boosters" over the next couple of years. Despite each session taking just fifteen minutes to complete—the briefest segment of time in the school calendar—the self-affirmation exercises reduced the racial differences in individual exam grades by 40 percent.[45] Even more amazingly, the effects could still be seen up to nine years after the original intervention. Overall, 92 percent of the Black children who had engaged in self-affirmation enrolled in college, compared with 78 percent of the Black children in the control group.[46]

Such large, long-term effects suggest a virtuous cycle of change. Bolstering feelings of self-worth against others' more negative expectations immediately boosts performance, which can then build personal confidence for later tests. With time, the act of appraising their values can help someone to reject society's self-fulfilling prophecies, creating a trajectory that is radically different from the prescribed path.[47]

Self-affirmation is now one of the most reliably tried-and-tested interventions to combat the effects of negative stereotypes.[48] If we are looking for new ways to reduce academic inequality, its widespread use is really a no-brainer.

Jacobson and Rosenthal concluded their "Pygmalion" study of Spruce Elementary with some words from the George Bernard Shaw play, as Eliza Doolittle describes the effects of others' expectations: "You see, really and truly, apart from the things anyone can pick up (the dressing and the proper way of speaking, and so on) the difference between a lady and a flower girl is not how she behaves, but how she's treated. I shall always be a flower girl to Professor Higgins, because he always treats me as a flower girl, and always will; but I know I can be a lady to you, because you always treat me as a lady, and always will." (They even describe it as "Shaw's summary" of the effect.)

The bittersweet tone of Doolittle's words did not reflect Jacobson and Rosenthal's optimism. Their writing at the time fizzed with excitement at the possibility that we would soon be able to raise children's abilities through a deeper understanding of the Pygmalion effect. Six decades later, though, we are still only just getting to that starting point.

The delay is frustrating, but we should not be surprised by how long it has taken to finally reach that point. As we have seen throughout this book, accepting the expectation effect requires us to overturn many of our assumptions about the brain, the body, and society—and this reversal of our beliefs necessarily requires a large body of evidence. With the recent resurgence of interest in the expectation effect, however, we finally have the knowledge and understanding to tap our own—and others'—potential through the extraordinary power of the Pygmalion phenomenon.

If all the world's a stage, then our scripts are often written by the people around us. In the past we may have unknowingly played those roles like unwitting cast members. But it needn't be that way. By learning to recognize the scripts we have been assigned, we can decide to reject the narratives that do not suit us—and create our own destinies.

## HOW TO THINK ABOUT . . . INTELLIGENCE, LEARNING, AND CREATIVITY

- Try to honestly assess your abilities and question whether you have internalized negative expectations. Is there really good reason to think you are inherently bad at math or art, for instance? Or might you have the capacity to improve?
- Once you have identified potential areas of growth, try to test whether those negative assumptions are true by looking for new challenges that push you out of your intellectual or creative comfort zone.
- Throughout this process, recognize that any moments of frustration are themselves a sign of effective learning and reflect the importance of the task at hand. This simple reframing will, by itself, increase your performance.
- If you feel especially anxious or believe you may be suffering from stereotype threat, try to practice self-affirmation (pages 221–25). As a way of neutralizing your fears and negative expectations, note the many other personal characteristics or

values that matter to your identity, and the reasons that they are important to you.

- If you are a teacher or manager, try to think about the ways your behaviors may be transmitting your expectations to others, both verbally and nonverbally. You may not be conscious of your body language or tone of voice, so it could be helpful either to ask an outside observer to watch your interactions or to film yourself interacting with your colleagues.

# 10

# THE SUPER-AGERS

## Why you really are as young—or old—as you feel

For more than a decade, Paddy Jones has been wowing audiences across the world with her spicy salsa dancing. She came to fame on the Spanish talent show *Tú sí que vales* (You're worth it) in 2009, and has since found success in the UK, through *Britain's Got Talent*; in Germany, on *Das Supertalent*; in Argentina, on the dancing show *Bailando*; and in Italy, where she performed at the Sanremo Music Festival in 2018 alongside the band Lo Stato Sociale.[1]

Jones also happens to be in her mideighties, making her the world's oldest acrobatic salsa dancer, according to Guinness World Records. Growing up in the UK, Jones had been a keen dancer and had performed professionally before she married her husband, David, at twenty-two and had four children. In retirement, the pair moved to Spain, and it was the tragedy of her husband's death from cancer that spurred her to take dance lessons. After trying all kinds of Latin American styles, she soon fell in love with acrobatic salsa dancing, during which she is frequently thrown into the air by her dance partner, Nico. "I don't plead my age because I don't feel 80, or act it," Jones told the media in 2014. She will only stop dancing, she has said, when Nico—forty years her junior—gets tired.[2]

We have now seen the many ways in which our expectations can powerfully influence our mental and physical well-being—altering our perception; our biological responses to diet, exercise, and stress;

and our cognitive abilities. And now I want to show you how all these expectation effects may powerfully converge to change the way we age. The fact that your beliefs could add or subtract years from your life is, in my opinion, the most striking and important consequence of this new understanding of the brain and the brain's prediction machine—and the reason I believe that it needs to be taken so seriously.

Before we continue, please give honest answers to the following four questions:

1. Do things get better, worse, or stay the same as you get older?
2. In each pair, which of the following words do you associate with retirement and beyond: uninvolved or involved; unable or able; dependent or independent; idle or busy?
3. When does middle age end and old age begin?
4. Based purely on your subjective experience (rather than your actual chronological age), how old do you feel today?

As we shall now see, your responses to these and similar questions may be as—or more—important for your future health than your current health status. Indeed, many scientists are coming to the conclusion that your beliefs about the aging process may be almost as important for your long-term well-being as your *actual* age.[3] Through multiple pathways, your expectations set the speed of your cells' biological clock, determining everything from trivial aches and pains to your risk of heart disease, dementia, and death. The youthful mindset of someone like Paddy Jones is, it turns out, a kind of elixir of youth.

## IF YOU COULD TURN BACK TIME . . .

The first hints that our thoughts and expectations could either accelerate or decelerate the aging process came from a remarkable experiment by the psychologist Ellen Langer at Harvard University.

Langer is known for being something of a maverick researcher; she was one of the first researchers to examine the benefits of mindfulness, long before the subject became a fashionable object of scientific study. (She was also the researcher who examined the ways that expectation can influence eyesight, as mentioned in Chapter 1.) In 1979 she decided to investigate the mind-body connection by asking a group of seventy- and eighty-year-olds to pretend that they were living in the year 1959 again.

The participants were recruited through local newspaper ads, and first they were given various tests that would normally be used to diagnose age-related problems. They were asked to perform a memory test along with other cognitive tasks, such as finding their way through pencil-and-paper mazes, that were designed to measure the speed of the brain's processing, which is commonly assumed to slow down in old age. Langer's team also tested the participants' eyesight and hearing, and the flexibility of their joints.

The researchers then brought the participants to a weeklong retreat at a monastery in Peterborough, New Hampshire, which had been redecorated to look as if it had been stuck in a time warp from the late 1950s. Everything, from the magazines in the living room to the music playing on the radio (such as the crooning of Perry Como, Nat King Cole, and Rosemary Clooney) and the films available to watch (*Some Like It Hot*, *North by Northwest*, and *Ben-Hur*), was carefully chosen for historical accuracy. To ensure that the environment changed the participants' mindsets, the researchers also asked them to write a biography of themselves for that era in the present tense, giving them explicit instructions to live as if it were 1959 without discussing anything that had occurred since that point. Instead, the participants were encouraged to discuss the politics and the sporting events that had occurred two decades previously. The aim was to evoke their younger and fitter selves through all these associations.

To create a comparison, the researchers ran a second retreat a week later. While factors like the décor, diet, and social contact remained the same, these participants were asked to reminisce about the past without being encouraged to act as if they were younger. When they wrote a biography describing their lives, they did so in the past tense, for instance, rather than the present tense—a seemingly small difference that meant their mindset was still focused on their current age.

Most of the participants showed some improvements in their final tests, compared to their baseline scores, but it was those in the first group, who had more fully immersed themselves in the world of 1959, who saw the greatest benefits. Sixty-three percent made a significant gain on the cognitive tests, for example, compared to just 44 percent in the control condition. Their vision became sharper, their joints more flexible, and their hands more dexterous, as some of the inflammation from their arthritis receded. The change was even noticeable in their appearance: as their posture improved they became taller and walked more easily. Langer took photos of the participants before and after the retreat; observers who were not told about the purpose of the experiment rated the second pictures as looking considerably younger than the first. It was as if Langer had actually turned the clock back.[4]

As enticing as these findings might seem, Langer's experiment suffered from the same flaws as some of the other early mindset studies (and indeed, much of the other psychological research from that era). Most serious was the size of the sample. There were just eight members in the immersive group and eight in the control group, which is not typically considered to be a large enough sample to draw general conclusions about the population as a whole. Extraordinary claims need extraordinary evidence, after all—and the idea that our mindset could somehow influence our physical aging is about as extraordinary as scientific theories come.

Becca Levy, at the Yale School of Public Health, has been leading the way in providing sufficient evidence to support the astonishing claim. In one of her earliest—and most eye-catching—papers, she examined data from the Ohio Longitudinal Study of Aging and Retirement. The founders of the study had selected more than eleven hundred participants who had turned fifty by July 1, 1975, and then followed their progress over the subsequent decades. The participants had been asked at the study's start to rate their agreement with statements such as the following:

- I have as much pep as I did last year.
- As you get older, you are less useful.
- Things keep getting worse as I get older.

Based on these responses, Levy's team divided the participants into two groups—those with a positive perception of their own aging and those with a negative self-perception of aging, and then the researchers examined the risk of mortality in each case.

She found the average person with a more positive attitude to aging lived on for 22.6 years after the study commenced, while the average person with poorer perceptions of aging survived for just 15 years—a difference of around 7.5 years. This link remained even when other known risk factors, such as their socioeconomic status or their feelings of loneliness, were taken into account. The implications of the finding are as remarkable today as they were in 2002, when the study was first published. "If a previously unidentified virus was found to diminish life expectancy by over seven years, considerable effort would probably be devoted to identifying the cause and implementing a remedy," Levy and her team wrote in their paper. "In the present case, one of the likely causes is known: societally sanctioned denigration of the aged."[5]

Later studies have since reinforced the link between people's expectations and their physical aging, while dismissing some of the more obvious—and less interesting—explanations. You might expect that people's attitudes will reflect their decline rather than contribute to the degeneration, for example. Yet that theory can't fully explain the most eye-catching results. Levy looked at the Baltimore Longitudinal Study of Aging, for instance, which had followed the progress of hundreds of people from the late 1950s to the early twenty-first century. Beginning in 1968, the participants were questioned about their attitudes to old age—such as their level of agreement with the statement that "old people are helpless." With an average age of thirty-six, it is unlikely that most of the participants would have started to suffer serious age-related disabilities; their opinions of aging were much more likely to have come from the culture around them than from any personal experience. And Levy found that those views could predict their subsequent risk of illnesses such as angina, congestive heart failure, myocardial infarction, and stroke up to thirty-eight years later, even when she controlled for preexisting factors such as obesity, smoking habits, or a family history of cardiovascular disease.[6]

Positive attitudes to aging even seem to protect us from certain kinds of dementia. Although the precise causes of Alzheimer's disease are still being researched, we know many of the neurological changes that accompany the illness, including the buildup of a protein called beta-amyloid between cells. As these clumps—named plaques—accumulate, they destroy the synapses that are essential for brain signaling. Patients with Alzheimer's also develop tangles of another protein, tau, within the brain cells. We now know that certain gene variants—most notably APOE4—can render you more vulnerable to the disease. But those inherited differences do not seal your fate; many people with APOE4 never develop dementia.

To find our whether your attitudes to aging might change the odds of being struck with Alzheimer's disease, Levy looked again at the medical records of the people within the long-term cohort studies that had measured people's attitudes to aging, one of which had fortuitously included regular MRI scans during the study and brain autopsies after death. She found that people's expectations became inscribed all over their brain, with a markedly increased accumulation of the beta-amyloid plaques and tau protein tangles in those who had previously shown a negative view of aging. These people also showed pronounced damage to the hippocampus, the seahorse-shaped region deep in the brain that is responsible for memory formation.[7]

The effects of age beliefs on dementia incidence

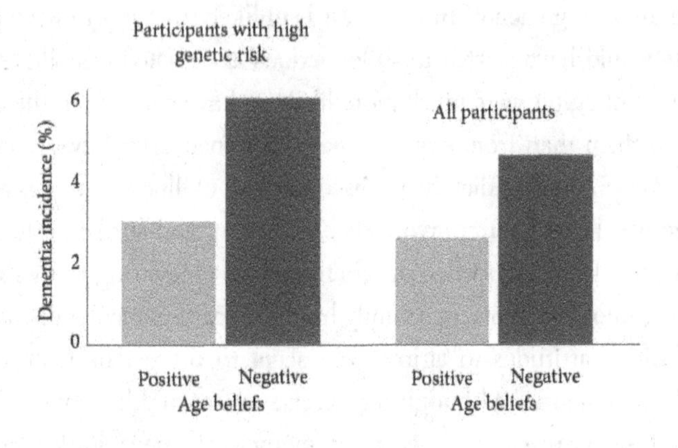


A follow-up study found that the effects of attitudes toward aging were particularly pronounced among the people carrying the high-risk ε4 variant of the APOE gene. For them, positive expectations of aging halved their risk of developing dementia, compared with people carrying the high-risk variant who assumed that aging came with a mental and physical decline. Indeed, among people with the positive expectations of aging, the high-risk variant of the gene barely seemed to increase the risk of dementia at all.[8]

It would be hard to overestimate the importance of these findings. So much has been said about purely biological risk factors that can accelerate the progression of disease as we get older. Yet according to this research, our own thoughts are equally if not more potent. High blood cholesterol levels, for example, are thought to reduce average life expectancy by up to four years—much less than the seven-and-a-half-year reduction caused by taking a dim view of our future health.[9]

Like any medical risk, the personal danger to ourselves of a negative attitude to aging will depend on many different factors. Even the simple question of how you define "old" could determine the extent of the effects—according to a seminal study of Whitehall civil servants working for the UK government.

The Whitehall studies are most famous for revealing the ways that social status can affect our health, showing that people on the lower rungs have a much greater health burden than those at the top of the competitive hierarchy. But in the early 1990s, civil servants were asked to define when middle age ended and old age began. And it turned out that the earlier they saw the onset of old age, the more likely they were to experience declining health at a younger age themselves. Over the following decade, people who believed that old age began at sixty or younger were around 40 percent more likely to develop coronary heart disease than those who believed that middle age finished at seventy or older.[10] In other words, it seems that you may be able to escape some of the effects by deciding that you haven't yet reached the relevant age bracket.

This brings us to the fourth and final question on page 229, in which I asked you to estimate your "subjective age"—how old you feel inside, as opposed to your actual chronological age. The idea that you're "as young as you feel" is something of a cliché, but studies examining thousands of participants have shown that people with a lower subjective age tend to enjoy greater physical and mental health.[11] How could this be? One possible answer is that the lower subjective age leads you to think that you are the exception to the normal decline that you expect in other people. This belief allows you to retain more positive expectations of your health as the years go by, shielding you from the damaging effects of the negative stereotypes that are normally so influential.[12]

This, essentially, is what Langer was trying to achieve in her timewarp study way back in 1979. In the monastery set up to replicate the décor and culture of the late 1950s, she hoped to turn back the participants' subjective ages by twenty years. When they had entered the house, they still felt like elderly people in their seventies and eighties, with all the perceived burden that this entailed. By the end, however, they left the monastery with the reinvigorated sense of themselves as fifty- and sixty-year-olds, ages when they had more energy and a greater sense of purpose in their lives. And at least for that small sample of participants, it seemed to have worked. Mentally—and, to a lesser but still significant extent, physically—they appeared to have temporarily turned back the clock.

## STEREOTYPE EMBODIMENT THEORY

Where do our negative expectations come from? And how do they wield such power over our well-being? To answer these questions, we need to understand a process known as "stereotype embodiment."

The writer Martin Amis provides a useful illustration. If the dancer Paddy Jones represents the optimum attitude to aging, then Amis offers us the polar opposite. In an interview in 2010, he decried the "silver tsunami" of the aging population. "There'll be a population of demented very old people, like an invasion of terrible immigrants, stinking out the restaurants and cafes and shops," he said. "I can imagine a sort of civil war between the old and the young in 10 or 15 years' time." He flippantly called for "euthanasia booths" on every street corner, and at literary festivals he described the aging process as starring in your own "low-budget horror film, saving the worst till last."[13] It is hard to imagine a harsher view of the elderly than to believe that death is preferable to healthy aging.

As literary critics pointed out at the time, Amis's novels had long expressed fear and disgust of aging, filled as they are with negative stereotypes of older generations. (The age of twenty, he claimed in his first novel, represents the end of youth.)[14] And as Amis has aged himself, he has often been filled with fear about his own fate. "Your youth evaporates in your early 40s when you look in the mirror," he told *Smithsonian* magazine. "And then it becomes a full-time job pretending you're not going to die."[15] In his sixties, he could already see his own talents dwindling, describing the loss of "energy and musicality" in his writing. The "waterfall" of creativity he had once felt was drying up.[16]

According to scientific research, Amis's experience represents a common trajectory. We pick up our negative views of the old in youth, when they are initially directed to other people. At a certain point, however, something changes in our own lives—we reach a milestone age, or we retire, or we go gray—leading us to realize the stereotypes now apply to us. At this point we begin to live out a self-fulfilling prophecy, as the stereotypes become "embodied," precipitating our physical and cognitive decline.[17]

There are multiple—simultaneous, but potentially interconnected—pathways to stereotype embodiment. The first is purely psychological. Consider the apparent loss of memory. When presented with negative stereotypes, older people tend to lose confidence in their mental abilities and prefer instead to rely on artificial crutches—like shopping lists or the car GPS—rather than committing things to memory. Yet research suggests that they can often remember far more than they suspect if they are forced to rely on their memory alone, and that making a concerted effort to exercise their minds should slow the decline.[18]

Problems with concentration can similarly arise from negative expectations; the more someone fears being distracted and proving the negative stereotype, the harder it is to focus. For many people, the shrinking attention span is an illusion that does not necessarily reflect a biological reality.[19] Gerben Westerhof at the University of Twente in the Netherlands has shown that something as inconsequential as viewing an ageist TV ad can impair people's thinking: seeing an older adult acting incompetently leads older viewers to suffer from impaired cognition. Such expectation effects may start out being temporary, but over time they could become ingrained, leading to more permanent decline.[20]

The second pathway is behavioral and motivational. If we assume that our body is set to become feeble and weak—and we see our environment as being more daunting than it really is—we will be discouraged from hard exercise and may find any exercise we do engage in as much more physically tiring thanks to our negative expectations. Even everyday movements—like someone's walking pace—become slower and less energetic when people are expecting a steep decline.[21] This may explain why Levy has found a strong correlation between people's attitudes to aging and their risk of obesity as they got older.[22]

Third, and finally, we have the psychosomatic route to becoming the stereotype we fear. Expectations of frailty can amplify the body's aches and pains or increase feelings like nausea and dizziness—a nocebo response that could contribute to the general perception of "feeling unwell"[23] that many older people report. Through changes to our respiration and metabolism, physical activity may become more difficult for us. (We have seen, after all, how people who take a dim view of their fitness tend to have a harder time exercising and are less likely to gain much benefit from it.)

Even more important, our negative expectations may provoke an unhealthy stress response, with far-reaching effects on our long-term health. Remember that the prediction machine carefully weighs up our abilities to respond to a new threat or challenge and uses these calculations to calibrate the release of hormones such as adrenaline and cortisol (which prepare us to deal with an immediate threat at the expense of long-term health) and DHEAS, which is involved in tissue maintenance and repair, and which is released in greater quantities when we see an event as a positive challenge. The prediction machine also controls the cardiovascular response, determining whether our vessels constrict (to prevent blood loss in the face of a threat) or dilate (to allow the brain and limbs to become oxygenated, allowing us to rise to a challenge), and whether we need to preserve energy or can release our stores to confront the situation. If someone considers themselves to be weaker, dimmer, and more vulnerable as a result of their age, they are more likely to see a difficulty as a negative threat rather than a positive challenge—resulting in more damaging stress responses that could wreak havoc on the body over time.

This has been apparent in laboratory experiments. Elderly people who have been primed with negative age stereotypes tend to have higher systolic blood pressure in response to stressful challenges, while those who have seen positive stereotypes demonstrate a more muted reaction.[24] In the long term, Levy has found that people's cortisol levels steadily rise by around 40 percent from around the age of fifty to around the age of eighty if they have negative attitudes to aging. Those with positive views, in contrast, show a 10 percent decrease in cortisol over the same period as they settle into the next stage of their life.[25] The chronic stress response can then trigger chronic inflammation, which causes general wear and tear on our tissues and is a known contributor to various illnesses, including arthritis, heart disease, and Alzheimer's. Sure enough, Levy has recently shown that negative attitudes to aging could predict heightened inflammation four years later, which, in turn, contributed to an increased risk of death over the following two years.[26]

The consequences of our negative expectations can even be seen within the nuclei of the individual cells, where our genetic blueprint is stored. Our genes are wrapped tightly in each cell's chromosomes, which have tiny protective caps—called telomeres—that keep the DNA stable and stop it from becoming frayed and damaged. (For this reason, telomeres are often compared to aglets, the plastic tips at the ends of shoelaces—a comparison that may be technically apt but nevertheless lacks a certain poetry, given just how important our telomeres are to our survival.) At birth our telomeres are long and strong, but they can be worn down by chronic stress and become shorter over our lifespan. Shorter telomeres reduce a cell's capacity to replicate without error, and without a long enough telomere, a cell may be unable to divide at all.[27]

The length of telomeres can vary among people of the same chronological age depending on lifestyle factors, including inflammation and stress, and this does seem to predict people's longevity and risk of disease. Levy's stereotype embodiment theory predicts that people with negative expectations of aging should therefore have shorter telomeres—and the evidence shows that this is indeed the case.[28]

By triggering unhealthy stress responses and promoting inflammation, our attitudes to aging can also affect the expression of the individual genes tucked inside those chromosomes. Within each cell we have small attachments to the DNA that can either "turn on" or "turn off" individual genes. This "epigenetic modification" determines which proteins the cell produces and ultimately how it functions. Certain patterns of activation or deactivation become more prevalent as we age and may explain many of the changes associated with old age, including our increased vulnerability to diseases. Importantly, people with negative attitudes to old age demonstrate more of these characteristic age-related changes, while people with more positive attitudes have a slower "epigenetic clock."[29]

Levy speculates that this may explain why someone's views of aging can determine the effects of the APOE gene on their risk of dementia. Worse stress may lead to epigenetic changes that heighten the gene's effects in the people with the negative views—increasing their vulnerability to the disease. People with a positive view of aging will lack that trigger, so the gene remains less active and has a smaller influence over their health.[30]

Some of these changes may be reversible.[31] We know that an enzyme called telomerase can help to repair the telomere caps at the ends of our chromosomes—and activating that enzyme seems to reverse some of the effects of premature aging. Perhaps in the future we will find drugs that prevent our cells from gathering any wear and tear. For now it seems certain that we can at least slow our decline by living a healthier lifestyle and by changing our expectations of what it means to be old.

## AGE AIN'T NOTHING BUT A NUMBER?

If we want to reappraise the true limits that come with age, let's first meet some more people who, like Paddy Jones, have confounded society's ageist expectations of what we can achieve later in life.

Take Hiromu Inada from Chiba, Japan. He started swimming, running, and cycling eighteen years ago and entered his first triathlon a year later. The sport became something of an obsession, and he eventually graduated to the Hawaii Ironman event in Kailua-Kona—an extreme endurance competition that requires the entrants to swim 2.4 miles, cycle 112 miles, and then run a full 26.2-mile marathon. To be able to do so, Inada developed a ruthless training schedule. He wakes at four-thirty each morning and hits the gym at six a.m. His training often continues until after sunset, and he takes only one day off each week.

By 2020 Inada had completed the Ironman three times, with finishing times that hover around 16 hours and 50 minutes. Simply making it to the finish line of an Ironman tournament would be a remarkable achievement for most people of any age. But Inada only began his training when he retired from his job as a news reporter in his sixties. He competed in his first Olympic-distance triathlon a few years later and began competing in the Ironman events when he was in his early eighties. His latest record came in 2018, a little over a month away from his eighty-sixth birthday.

As the work on subjective aging would predict, Inada has maintained a youthful outlook, and he does not view age as a barrier to extraordinary achievement. He says that even at seventy he felt "very young," and his training has helped to stave off a decline since then.

Equally impressive is the Swiss runner Albert Stricker, who completed an ultramarathon in Basel in Switzerland at the age of ninety-five. Like Inada, he only started his sport after retirement, aged sixty-five, and ran his first full marathon aged ninety. His training consists of runs of up to six miles every weekday. The aim, in the Basel event, was to run as far as you could for twelve hours solid; Stricker covered around 33 miles in total. You may imagine that this kind of exertion would cause serious bodily damage to someone of his age, but medical tests by Beat Knechtle at the Institute of Primary Care, University of Zurich, revealed that Stricker had completely recovered from the event within five days.[32]

It's fair to say that Jones, Inada, and Stricker will not be troubling the records of elite younger athletes anytime soon; all other things being equal, younger people will have a physical advantage. But this trio nevertheless demonstrates that an extraordinarily high level of fitness—even extreme endurance—can be achieved later in life.

Knechtle's analyses of male ultramarathon runners, for example, found that there's around an 8 percent drop in performance for each decade lived.[33] And even these age-related differences may be shrinking as older athletes become more competitive and find better ways to maintain their fitness. In the 1980s, for example, the sixty- to sixty-four-year-olds were competing at around 60 percent the capacity of the under-forties in Ironman tournaments; now it's more than 70 percent.[34] The truth is, we still don't know how well we can preserve our fitness into old age—since so few people have tried to push their bodies to these limits. But we have more than enough evidence to show that our potential as we age is much greater than is commonly believed. These patterns match studies of less extreme exercisers, which show that—when you treat your body correctly and if you adopt the right lifestyle *and* the right mindset—the human body can be much more resilient to the passage of time than most of us imagine.

As we appraise the cognitive effects of aging, we should remember the astonishing bursts of creativity shown by certain artists as they reached the average retirement age. Consider Penelope Fitzgerald. Having worked in various jobs, including teaching, she published her first novel at the age of sixty and won the Booker Prize two years later. At the age of eighty she won the US National Book Critics Circle Award for her final novel, *The Blue Flower*—a book often considered to be her masterpiece.[35] "You can see her becoming a better writer as she aged," wrote the *New Yorker* critic James Wood, "more serious and expansive, more confident and supple."[36] It's hardly the creative drought that Martin Amis has described.

In the visual arts, Pablo Picasso and Henri Matisse both found renewed inspiration toward the end of their lives. At around the age of sixty, Picasso turned to ceramics, creating more than 3,500 works that fused painting, printmaking, and sculpture.[37] Matisse, meanwhile, picked up scissors and paper to produce his astonishing "cut-outs," which he described as "carving into colour." They remain some of his most celebrated works.[38]

These extraordinary stories are worth bearing in mind, since a recognition of our capacity to control our aging, be it through mindset or lifestyle changes, can itself provide a kind of counter-charm against the negative stereotypes of getting older.

Consider a study of sixty- to ninety-year-olds by David Weiss at Columbia University. In 2018, he first tested whether his participants endorsed essentialist views such as the following:

> To a large extent, a person's age biologically determines his or her abilities.

Or whether they considered aging to be a malleable process, endorsing statements such as the following:

> No matter at what point you are in your life, you can always influence your own aging.

Note that these beliefs, in themselves, do not necessarily reflect a "good" or "bad" view of aging—but whether the process is controllable.

After they had completed the survey, Weiss asked the participants to take a quiz that consisted of questions about dementia and physical disability—an exercise designed to evoke the typical stereotypes of older people. Finally, he asked them to complete a memory test and measured the stress response. He found that the people who saw aging as a biological inevitability were far more likely to be affected by the negative stereotypes in the dementia quiz, showing greater stress and reduced performance in the memory test. That makes sense: if you assume that you have no control over your biology, then the thoughts of decline are going to be much scarier.

The people who felt that they had more control over their fate were not adversely affected by the fears of aging. If anything, they tended to do better on the test after being primed with all those thoughts of frailty and decline than people who had not been exposed to the negative stereotypes.[39] They seemed to have been energized by the opportunity to prove themselves as exceptions to the pessimistic predictions of aging.

You don't need to have the ambition to be an Ironman athlete, award-winning novelist, or prolific artist to understand that aging can be a far more exciting prospect than the doom and gloom depicted by the likes of Martin Amis; their stories simply show us the extraordinary limits of what is possible. The way we age is very much within our power, and the more we remember that, the easier we will find it to confront the negative expectations thrust on us by society, and to choose our own path. As Hiromu Inada told the *Japan Times*: "I hope everyone can see and be encouraged that you can do the same things as the younger generation."[40]

## ELIXIRS OF LIFE

A broader recognition of these age-related expectation effects cannot come quickly enough. In 2015 there were around 901 million people aged sixty and above—12.3 percent of the global population. By 2030 that number will have risen to 1.4 billion (16.4 percent of the global population), and by 2050 it will have increased to 2.1 billion (21.3 percent of the global population).[41] At current rates of diagnosis, 152 million of those people may have dementia by the middle of this century.[42]

Doctors today often talk about the health span (those years lived without serious disability or illness) as opposed to the life span—the idea being that living a good life without illness is the true goal, rather than simply extending the number of years you survive.[43] But by raising our expectations of the aging process, we have the astonishing possibility of adding years to both. It is little wonder that scientists have been investigating the best ways to apply this research on a large scale.

As part of her continuing research on the effects of ageism, Yale's Becca Levy invited elderly participants—aged between sixty-one and ninety-nine—to play a computer game while positive age-related words (such as "wise," "spry," and "creative") flashed briefly on the screen. Although the participants could not have perceived the words consciously, they must have absorbed the message, as Levy found that their attitudes toward aging improved significantly over the course of four weekly sessions. And this newfound optimism translated to a remarkable improvement in their physical well-being: they were more mobile, and their gait and posture had begun to resemble those of a younger person. Amazingly, these benefits—gained from implicit messaging—even surpassed the results of a physical exercise regimen that encouraged gentle activity three times a week for six months.[44]

Levy's experiment was an important proof of concept to demonstrate how powerful unconscious cues can be in changing people's expectations, and the benefits they can bring. Given these results, some researchers have speculated that similar cues could be added to certain films or TV programs, though such messaging would have to contend with some people's discomfort at the idea of subliminal manipulation.[45]

For now, it may be more realistic to focus on conscious change, without the use of any hidden cues. The most exciting interventions combine education about aging stereotypes with other activities, like physical exercise, allowing people to test their own abilities and gain some firsthand evidence of the ways that their expectations may be limiting their lives.

The benefits have been remarkable. Elderly residents of Los Angeles, for instance, were given weekly lessons on the physical potential of the aging body and brain, and on the ways that negative stereotypes might be holding them back, followed by an hour's gym class to reinforce the learning. After the lessons had finished, their mobility had increased tremendously—from 24,749 to 30,707 steps per week, a 24 percent increase, according to the readings on their pedometers. Crucially, the physical benefits seemed to follow changes to their attitudes to aging—the more positive their thinking, the more active they became. The participants also reported better day-to-day functioning, and reduced pain from chronic conditions (like arthritis).[46]

These results have since been replicated many times with many populations. In some cases, the altered expectations were found to double the participants' physical activity long after the intervention had finished, with improvements above and beyond standard fitness regimens that had not deliberately targeted participants' expectations about aging.[47] While it is hard to pick apart the precise reasons for such improvement, it seems likely that the intervention acted on all three elements of "stereotype embodiment"—the psychological, behavioral, and psychosomatic factors. The participants' positive expectations reduced age-related stress and improved the way they felt, physically and mentally, and this, in turn, meant that they were more likely to exercise.

Ideally, these kinds of interventions will soon be provided by health services across the world. In the meantime, we can all start applying a bit of critical thinking to our own thoughts. If you feel that you may be too old for a certain activity, start questioning the foundations for that expectation. Does it arise from a real physical disability that you actually feel at this moment? Or have you been infected by the messages of others? And is it time to push yourself out of your comfort zone with a new activity you were once too scared to try? When it comes to our cognitive decline, there is now good evidence that learning new skills in middle and old age can help to maintain memory and concentration, and, crucially, build your confidence in your own abilities, reversing some of your more negative expectations and sparking a virtuous cycle.[48]

When I spoke to Paddy Jones, she was careful to emphasize the potential role of luck in her good health. But she agrees that many people have needlessly pessimistic views of their capabilities during what could be their golden years, and she encourages them to question those views. Since she came to fame, she has received many messages from people who have been inspired to take up new activities, and she hopes that others will follow their lead. "If you feel that there's something that you want to do, and it inspires you, try it! And if you find you can't do it, then look for something else that you can achieve."

Reevaluating your attitudes to age and aging will be especially important when you face a serious life event like retirement. Paddy Jones and the ultra-endurance athletes Hiromu Inada and Albert Stricker all took up their sports after finishing their earlier careers. At a time when many people start thinking more pessimistically about their aging, they found a way to challenge those attitudes and continually prove their own abilities—and we could all learn from their experiences, however grand or modest our own ambitions may be.

## AN AGELESS SOCIETY?

In November 2019 I was lucky enough to visit the province of Nuoro, on the east coast of Sardinia. Its rugged mountains rise steeply from the Mediterranean Sea, and its 200,000 inhabitants live in small villages and towns dotted throughout the valleys.[49] Farming—of goats and pigs—is still the main way of living.

In the past, Nuoro was most famous for being the birthplace of the Nobel Prize–winning writer Grazia Deledda. Today it is perhaps best known for having one of the highest concentrations of centenarians in the world. When adjusted for the size of the overall population, the population of Nuoro has about three times as many people over the age of a hundred as in the rest of Sardinia, and ten times as many as in the USA.[50]

There are many scientific theories for their incredible longevity. The population of Nuoro has been isolated for large periods of history and has a unique genetic profile as a result. As we've already seen, however, our genes cannot seal our fate; a study from 2018 found that just 7 percent of the differences in longevity can be attributed to our genes.[51] There's also the diet in Nuoro, which is spartan but nutritious, and high in antioxidants that are known to prevent cellular damage; and there is exercise, with some farmers continuing to work into their seventies and eighties.

Considering our knowledge of expectation effects and their power over our lives, however, I can't help but wonder whether a large part of the amazing life spans of people in Nuoro is down to their culture, which holds great respect for the older members of the community. Dr. Raffaele Sestu, a family doctor in the small town of Arzana, certainly thinks so. He has worked with dozens of centenarians in his practice, and he says that most are treated with reverence, as the head of the family, well into their old age. "Someone who knows they have a role, and who believes in him- or herself, lives a better life, and more easily lives longer than one hundred," he told me.[52]

Sadly, this attitude seems to be missing in many industrialized countries across the Americas, Europe, and Asia,[53] where fewer and fewer people live in intergenerational households, and older people are more often treated as a burden rather than a valued member of the family. This mindset is a disadvantage for children and grandchildren as well as grandparents: various studies show that regular contact with elders can lead younger people to develop more positive views of aging. As those children reach adulthood and middle age, those experiences will help them to remember what healthy aging can look like. People who do not have that regular contact, in contrast, are more easily swayed by the ageist stereotypes in the media.[54] When you do not regularly see people of a certain demographic, it can be easy to either ridicule or disparage them. It is a sad irony that our medical care has been able to increase life expectancy, and yet—due to other social changes—we have come to view those resilient golden-agers as a nuisance rather than people to be cherished and respected.

Places like Nuoro—where age is seen as a strength—might be much rarer today than they were in the past, but it needn't be this way. On a personal level we might try to build bridges between generations, befriending people who are older and younger than us. But as a society, we need to go much further and tackle ageism just as we do racism, homophobia, and other kinds of prejudice. Every time we lazily use ageist stereotypes, we are effectively spreading a deadly pathogen that will, in time, come to hurt ourselves as much as others.

When it comes to the expectation effects surrounding aging, we all have the choice to either perpetuate these toxic ideas or help to change them. And we need to act now; our lives, and those of the people we love, may literally depend on it.

## HOW TO THINK ABOUT . . . AGING

- Rather than idealizing youthfulness, focus on all the things that you can gain from living a longer life—including experience, knowledge, and improved emotional regulation and decision-making.
- Remember that many of the things that we typically associate with aging—such as physical weakness—are within your control and can be improved with a healthier lifestyle.
- Avoid attributing sickness to your age, since this will reinforce the idea of an inevitable decline. People with a positive view of aging tend to recover from illness more quickly than those with negative expectations.
- Look out for good role models—people like Paddy Jones or Hiromu Inada who have challenged society's expectations.
- Be aware of your media diet—many films and TV series will reinforce offensive stereotypes about older people. Try to watch stories or documentaries that deal with aging more sensitively—or at least engage more critically with what you do watch.
- If you are young or middle-aged, befriend people outside of your own age group—the research shows that doing so, on its own, can improve expectations of aging.

# EPILOGUE

Let's return to the Hmong in the United States and the "sudden unexpected nocturnal deaths" at the hands of the dab tsog. During the 1980s—at the very worst point of the crisis—it seemed incredible that believing in your impending doom could increase your risk of mortality. Yet that harmful prediction is perfectly in keeping with the cutting-edge recognition of expectation effects, and their power, that has emerged in the twenty-first century.

Inspired by this research, some doctors have started to take action. The Mercy Medical Center in California, for instance, actively works with shamans to improve the treatment of the area's large Hmong community.

It started with a single case study, when a Hmong man appeared to be dying from a gangrenous bowel. None of the treatments he was offered seemed to be taking effect, but well-wishers from the community asked the staff to allow a Hmong healer to help. The hospital eventually relented, and the shaman performed his rituals—including placing a sword above the ward door to scare off evil spirits. Despite his initial prognosis, the man subsequently made a full recovery and returned to being an active member of the local Hmong community.

"Physicians experience these 'miracles' from time to time," a spokesman from the Mercy Medical Center explained. "But this case really illustrated to them the power of these ceremonies. Healing isn't just about medicine, it's about people." The medical center has subsequently trained 140 shamans to work alongside doctors in the hospital, supporting the standard medical procedures with their rituals. The policy has encouraged more people to seek treatment from the hospital and—anecdotally at least—has improved the way the patients respond to medical cures.[1]

I hope that after you read this book it will be clear that we are all shaped by our beliefs in such ways. While these kinds of events may seem miraculous, they are astonishingly commonplace for people of every faith—or those with no religion at all.

Whether we are undergoing surgery, protecting our health and fitness, coping with prolonged stress, or working under enormous pressure, our expectations will shape our psychological and physiological responses to our circumstances. The brain evolved to make predictions, drawing on our own previous experiences, our observations of others, and our cultural norms—a process that underlies our very perception of reality and prepares the mind and the body for whatever we have to face. And we now know the ways we can reappraise those expectations to create our own self-fulfilling prophecies.

Throughout these chapters, I have tried to be clear that the growing recognition of expectation effects in no way downplays the enormous challenges our society as a whole is facing today and will no doubt face in the future. We can't just wish away financial uncertainty or social injustice: expectation effects are not a cure-all for every problem we encounter. Knowledge of them can, however, be a useful tool to build our own personal resilience, and occasionally the use of expectation effects may even allow us to thrive despite the difficulties, equipping us with the strength to bring about real change.

Ideally, practicing these skills should become a habit—so that whatever we are doing, and whatever new message we encounter, we probe and question the framing to see if we are accidentally forming a negative self-fulfilling prophecy with no rational basis. I've certainly found that doing so has changed my life, since I first learned about the nocebo origins of my antidepressant side effects. Knowledge of the expectation effect has changed the way I eat and exercise, my attitudes to sleep, and my thoughts about aging. Much of this book was written during the Covid-19 pandemic, and I found that the techniques I describe here were often invaluable in helping me to cope with the loneliness and stresses of the continual lockdowns.

I hope that you will find this groundbreaking understanding of the brain to be equally fruitful in your own life. You may have already seen some benefits; knowledge is power, and simply reading about the science of the expectation effect and its consequences can shift your mindset and have a measurable impact. If, however, you find that you are struggling to apply certain elements of this research, you might consider the following three strategies to gently nudge you out of your rut. Like all the other advice in this book, these last techniques are inspired by robust scientific evidence, and taken together they address some of the most commonly encountered problems.

### 1. REMEMBER THAT YOUR MIND IS A "WORK IN PROGRESS"

Let's begin with the idea of neuroplasticity—the brain's ability to rewire and change—which can itself be subject to an expectation effect.

In the early days of neuroscience, brains were thought to be static entities. While children's minds might be pliable—up to a point—the capacity for neural change was meant to disappear after adolescence, making it much harder to alter our abilities and personality traits. "In adult centres the nerve paths are something fixed, ended, immutable," the founder of modern neuroscience, Santiago Ramón y Cajal, wrote in 1928.[2] That would indeed be bad news for undoing your existing thinking habits. Whenever I discuss expectation effects, some skeptics ask whether we are "wired" to see the world in a certain way, and whether certain expectations are too deeply ingrained to change.

Thankfully, we now know that there is very little reason to be so pessimistic about our capacity for self-transformation. Through meticulous research, neuroscientists have shown that the brain's wiring is constantly changing—it strengthens some connections and prunes others and sometimes adds whole new networks in response to your circumstances. Those connections will then determine your abilities. At its most extreme, this process allows people who were born deaf or blind to adapt to cochlear or retinal implants; while their brains are initially unable to make sense of the new information, they soon rewire to build sounds and images. But neuroplasticity also occurs whenever we learn a new skill. Even some personality traits such as neuroticism or introversion, which were once thought to be completely immovable, can change over a lifetime.

Whatever your current situation, your brain may be much more malleable than you might think. And making a change will be much easier if you hold certain attitudes. Carol Dweck at Stanford University has found that some people believe their abilities to be fixed and immutable: either they are good at something or not. Others believe in their capacity for improvement, no matter what their initial aptitude. In general, people with the "growth mindset" tend to progress more quickly than people with a "fixed mindset."

The growth mindset is well known in education, but it is now becoming apparent that people's understanding of the brain's inherent malleability can have far-reaching consequences for many other kinds of personal change. People with anxiety or depression are more likely to benefit from treatments such as cognitive behavioral therapy if they have a growth mindset, for example, than if they have a fixed mindset.[3] Given these kinds of results, researchers are now looking at interventions that encourage the growth mindset in a range of settings. They have found that teaching people about the brain's capacity to change can itself improve people's physical and mental health, as people realize that they do not need to become stuck in their current thinking habits.[4]

If you find you are stuck in a rut as you try to apply a particular expectation effect, and you are struggling to reframe events in a more productive or positive way, you should try to remind yourself of the brain's plasticity. Rather than assuming that you are destined to fall into the same traps again and again, picture your brain rewiring as you learn to see the world in a new way. Since it's much easier to believe in the growth mindset when you have already experienced change, you may also find that it helps to focus on small, achievable goals that can prove your capacity for personal transformation, before steadily increasing your ambitions, and try along the way to view any failures as a useful learning experience.

You have had a whole lifetime to build your current worldview, after all, so it is only natural that positive change will take time. In the words of one research team studying the growth mindset: "Everyone's brain is a work in progress!"[5]

## 2. TAKE AN OUTSIDE VIEW

Even if you have a growth mindset, you may sometimes find it hard to apply an expectation effect in the heat of a particularly challenging moment. Reframing your pain, anxiety, or fatigue can sound easy in theory—but it's much harder when you're already in discomfort and struggling to hold yourself together.

In these situations, the first thing to remember is that you don't have to ignore those uncomfortable feelings, a feat that would be extraordinarily difficult to achieve and counterproductive. You can create a positive expectation effect by adjusting your assumptions about the feelings' meaning and consequences, rather than immediately changing the feelings themselves. You can remind yourself that your physical symptoms are a sign of the body's healing, for example, without actively suppressing the actual sensation of pain; similarly, you can remember the fact that anxiety is energizing, while still feeling stressed. In both cases, the shift in thinking should lead to healthier responses—without needing to deny, swallow, or change the feelings themselves.

To make that process of reframing easier, you might also try a technique known as "self-distancing," developed by Ethan Kross, a psychologist at the University of Michigan. According to Kross's research, our emotions are often too immediate for us to think objectively about our situation; instead they drag us into negative rumination, churning over the same fearful or unhappy thoughts that will, in turn, make us feel worse and less rational. If we force ourselves to take an outside perspective of the situation, however, he argues, we can all put an end to that negative ruminative cycle.

There are many ways to self-distance. You can imagine looking back at your current event from some time in the future, months or years away. Or you can imagine that you are an observer, watching the situation unfold from outside your body. The technique I personally find most useful is to imagine that I am advising a friend in the same situation.

There is now a wealth of evidence showing that these self-distancing strategies can gently relieve people's distress in a range of situations, which in turn allows them to reframe the situation more constructively. When facing a stressful event like public speaking, for example, people who self-distance are more likely to view the event as a positive challenge and a chance to prove themselves, rather than seeing it as a potential threat that could lead to embarrassment and failure.[6] As we've seen, that kind of mental shift pushes the body toward a healthier stress response.[7]

This public speaking example is just one of many in which self-distancing has been proven to shift thinking from negative rumination to more constructive reappraisals of the situation at hand—making it an incredibly useful tool for personal transformation. If I were trying to reframe my sense of pain from an illness, for example, I might try to think how I would reassure a friend in distress, reminding them, for instance, about the very real chances of recovery and the benefits of their treatment—thoughts that are much easier to express when you feel a little bit removed from the situation. The same goes for my thoughts about aging; I'm far less likely to build up a dismal view of my own future if I imagine that I am talking to someone else rather than myself. Instead, I'd be keen to emphasize all the opportunities that are still out there.

Whichever expectation effects you are trying to apply, a few moments of self-distancing should put you in a more constructive frame of mind so that you can more easily identify your preconceptions and adjust your beliefs to healthier ways of thinking.

### 3. BE KIND TO YOURSELF

My final piece of advice concerns your sense of responsibility. Knowledge of the prediction machine, and of our power to shape our responses to events through techniques like reframing, can be awe-inspiring. But there is also the danger that such awareness can create a sense of guilt or blame. If your nerves got the better of you during a speech, it's your fault for seeing stress as debilitating, you might think. If you're tired and can't face working another hour, it's because you've got the wrong mindset about willpower! If you're not as fit as you once were, you've been thinking yourself old!

These sentiments could not be further from my own views or those of the scientists exploring expectation effects—and the spread of these self-defeating ideas would be the very worst outcome to my mind. Like any tools, the strategies described in this book will suit some individuals better than others, and they will be more applicable to certain situations than to others. If you find that a particular technique doesn't work for you, move on and maybe—if you're feeling ready—return to it at a later date. The very last thing you want to do is to beat yourself up or imagine that your inability to change your mindset is a sign of personal failure.

Psychologists across the globe are coming to understand that an attitude of "self-compassion" is crucial for any personal transformation. This mindset involves acknowledging and accepting the many other factors that could contribute to your difficulties, and recognizing that many other people also share your difficulties; you are not alone in your struggles.

Self-compassion is in itself good for our mental and physical health—but equally important, it gives us a sense of safety that makes it much easier to forge new habits and effect positive change in our lives. And that includes the use of the reappraisal techniques that we've seen throughout this book.[8] The trick is to acknowledge your potential for improvement, without being overly critical—in much the same way you might advise a family member.

We should all adopt a self-compassionate attitude whenever we apply an expectation effect. The fact that we may have been holding unhealthy or harmful beliefs should be no source of shame, and there will inevitably be times when we struggle to shift our mindset. Like any skill, you need practice to bring about permanent change.

Whatever you hope to achieve with the expectation effect, try to keep an open mind when testing the different techniques, forgive any failures, and celebrate any successes. If you think that you are capable of personal transformation—and you are willing to forgive your mistakes—you can make that your self-fulfilling prophecy.

Shakespeare expressed it best more than four hundred years ago, with Hamlet declaring that "there is nothing either good or bad, but thinking makes it so." And with that realization, we can all take our fates into our own hands.

# NOTES

## Introduction

1. Crum, A.J., and Langer, E.J. (2007). Mind-set matters: Exercise and the placebo effect. *Psychological Science*, *18*(2), 165–71.
2. Sharpless, B.A., and Barber, J.P. (2011). Lifetime prevalence rates of sleep paralysis: A systematic review. *Sleep Medicine Reviews*, *15*(5), 311–15.
3. For a fascinating and in-depth discussion of the many factors that contributed to the Hmong deaths in the USA, see Adler, S.R. (2011). *Sleep Paralysis: Night-mares, Nocebos, and the Mind-Body Connection*. New Brunswick, NJ: Rutgers University Press.
4. Zheng, J., Zheng, D., Su, T., and Cheng, J. (2018). Sudden unexplained nocturnal death syndrome: The hundred years' enigma. *Journal of the American Heart Association*, *7*(5), e007837.
5. Alia Crum described the implications of mindsets at the World Economic Forum in January 2018: https://sparq.stanford.edu/sparq-health-director-crum-discusses-mindsets-world-economic-forum-video.

## Chapter 1: The Prediction Machine

1. These descriptions of the drone attacks are indebted to Shackle, S. (2020, December 1). The mystery of the Gatwick drone. *Guardian*. https://www.theguardian.com/uk-news/2020/dec/01/the-mystery-of-the-gatwick-drone. See also Jarvis, J. (2018, December 23). Gatwick drone latest. *Evening Standard*. https://www.standard.co.uk/news/uk/gatwick-drone-latest-police-say-it-is-a-possibility-there-was-never-a-drone-a4024626.html.
2. The term "prediction machine" was introduced by Professor Andy Clark (2016) in his book *Surfing Uncertainty: Prediction, Action, and the Embodied Mind*. Oxford: Oxford University Press. Others refer to it as the "prediction engine"—but for clarity and consistency, I shall use Clark's term throughout.
3. Von Helmholtz, H. (1925). *Treatise on Physiological Optics*, vol. 3, ed. James P.C. Southall, 1–37. Birmingham, AL: Optical Society of America. "It may often be rather hard to say how much of our apperceptions (Anschuuuwen) as derived by the sense of sight is due directly to sensation, and how much of them, on the other hand, is due to experience and training." See also Meyering, T.C. (1989). *Helmholtz's Theory of Unconscious Inferences*. In: T.C. Meyering (Ed.). *Historical Roots of Cognitive Science*, 181–208. Dordrecht: Springer.
4. Foa, M. (2015). *Georges Seurat: The Art of Vision*, 21. New Haven, CT: Yale University Press.
5. For an in-depth discussion of predictive coding and its many implications, see Clark, A. (2016). *Surfing Uncertainty: Prediction, Action, and the Embodied Mind*. Oxford: Oxford

University Press; Hohwy, J. (2013). *The Predictive Mind.* Oxford: Oxford University Press. See also De Lange, F.P., Heilbron, M., and Kok, P. (2018). How do expectations shape perception? *Trends in Cognitive Sciences, 22*(9), 764–79; O'Callaghan, C., Kveraga, K., Shine, J.M., Adams Jr, R.B., and Bar, M. (2017). Predictions penetrate perception: Converging insights from brain, behaviour and disorder. *Consciousness and Cognition, 47*, 63–74.

6. Barrett, L.F. (2017). *How Emotions Are Made: The Secret Life of the Brain*, 60. London: Pan Macmillan.
7. Fenske, M.J., Aminoff, E., Gronau, N., and Bar, M. (2006). Top-down facilitation of visual object recognition: Object-based and context-based contributions. *Progress in Brain Research, 155*, 3–21.
8. Bar, M., Kassam, K.S., Ghuman, A.S., Boshyan, J., Schmid, A.M., Dale, A.M., . . . and Halgren, E. (2006). Top-down facilitation of visual recognition. *Proceedings of the National Academy of Sciences, 103*(2), 449–54.
9. Madrigal, A. (2014, May 5). Things you cannot unsee. *Atlantic*. https://www.theatlantic.com/technology/archive/2014/05/10-things-you-cant-unsee-and-what-that-says-about-your-brain/361335.
10. Brugger, P., and Brugger, S. (1993). The Easter bunny in October: Is it disguised as a duck? *Perceptual and Motor Skills, 76*(2), 577–78. See the following for a discussion of this paper's interpretation in light of modern theories of predictive processing: Seriès, P., and Seitz, A. (2013). Learning what to expect (in visual perception). *Frontiers in Human Neuroscience, 7*, 668.
11. Liu, J., Li, J., Feng, L., Li, L., Tian, J., and Lee, K. (2014). Seeing Jesus in toast: Neural and behavioral correlates of face pareidolia. *Cortex, 53*, 60–77. See also Aru, J., Tulver, K., and Bachmann, T. (2018). It's all in your head: Expectations create illusory perception in a dual-task setup. *Consciousness and Cognition, 65*, 197–208; Barik, K., Jones, R., Bhattacharya, J., and Saha, G. (2019). Investigating the influence of prior expectation in face pareidolia using spatial pattern. In Tanveer M., and Pachori R. (Eds.). *Machine Intelligence and Signal Analysis*, 437–51. Singapore: Springer.
12. Merckelbach, H., and van de Ven, V. (2001). Another White Christmas: Fantasy proneness and reports of 'hallucinatory experiences' in undergraduate students. *Journal of Behavior Therapy and Experimental Psychiatry, 32*(3), 137–44; Crowe, S.F., Barot, J., Caldow, S., D'Aspromonte, J., Dell'Orso, J., Di Clemente, A., . . . and Sapega, S. (2011). The effect of caffeine and stress on auditory hallucinations in a non-clinical sample. *Personality and Individual Differences, 50*(5), 626–30.
13. As noted above, when we hallucinate something, the brain activity is very similar to the responses to actual physical images. Summerfield, C., Egner, T., Mangels, J., and Hirsch, J. (2006). Mistaking a house for a face: Neural correlates of misperception in healthy humans. *Cerebral Cortex, 16*(4), 500–508.
14. These details come from Huntford, R. (2000). *Scott and Amundsen: Their Race to the South Pole*, 567. London: Abacus.
15. Hartley-Parkinson, R. (2019, April 17). Mum claims she can see Jesus in flames of Notre Dame Cathedral. *Metro*. https://metro.co.uk/2019/04/17/mum-claims-can-see-jesus-flames-notre-dame-cathedral-9225760.
16. Dunning, D., and Balcetis, E. (2013). Wishful seeing: How preferences shape visual perception. *Current Directions in Psychological Science, 22*(1), 33–37. See also Balcetis, E. (2014). Wishful seeing. https://thepsychologist.bps.org.uk/volume-27/january-2014/wishful-seeing.
17. Greene, B. (2017). How does consciousness happen? https://blog.ted.com/how-does-consciousness-happen-anil-seth-speaks-at-ted2017.
18. https://rarediseases.org/rare-diseases/fnd/.

19. This case study is described in detail in the following paper: Yeo, J.M., Carson, A., and Stone, J. (2019). Seeing again: Treatment of functional visual loss. *Practical Neurology*, *19*(2), 168–72. Enormous thanks to Jon Stone for clarifying some details.
20. For a description of this kind of process, see Pezzulo, G. (2013). Why do you fear the bogeyman? An embodied predictive coding model of perceptual inference. *Cognitive, Affective, and Behavioral Neuroscience*, *14*(3), 902–11.
21. Teachman, B.A., Stefanucci, J.K., Clerkin, E.M., Cody, M.W., and Proffitt, D.R. (2008). A new mode of fear expression: Perceptual bias in height fear. *Emotion*, *8*(2), 296–301.
22. Vasey, M.W., Vilensky, M.R., Heath, J.H., Harbaugh, C.N., Buffington, A.G., and Fazio, R.H. (2012). It was as big as my head, I swear! Biased spider size estimation in spider phobia. *Journal of Anxiety Disorders*, *26*(1), 20–24; Basanovic, J., Dean, L., Riskind, J.H., and MacLeod, C. (2019). High spider-fearful and low spider-fearful individuals differentially perceive the speed of approaching, but not receding, spider stimuli. *Cognitive Therapy and Research*, *43*(2), 514–21.
23. Jolij, J., and Meurs, M. (2011). Music alters visual perception. *PLoS One*, *6*(4), e18861. See also Siegel, E.H., Wormwood, J.B., Quigley, K.S., and Barrett, L.F. (2018). Seeing what you feel: Affect drives visual perception of structurally neutral faces. *Psychological Science*, *29*(4), 496–503; Wormwood, J.B., Siegel, E.H., Kopec, J., Quigley, K.S., and Barrett, L.F. (2019). You are what I feel: A test of the affective realism hypothesis. *Emotion*, *19*(5), 788–98. "The present findings are consistent with recent empirical work demonstrating that one's affective state may influence how positive or negative a neutral target face looks to the perceiver in a very literal way (Siegel et al., 2018): neutral faces were perceived as looking more smiling when presented concurrent with suppressed affectively positive stimuli and as looking more scowling when presented concurrent with suppressed affectively negative stimuli." Otten, M., Seth, A.K., and Pinto, Y. (2017). A social Bayesian brain: How social knowledge can shape visual perception. *Brain and Cognition*, *112*, 69–77. O'Callaghan, C., Kveraga, K., Shine, J.M., Adams Jr, R.B., and Bar, M. (2016). Convergent evidence for top-down effects from the "predictive brain." *Behavioral and Brain Sciences*, *39*, e254.
24. Bangee, M., Harris, R.A., Bridges, N., Rotenberg, K.J., and Qualter, P. (2014). Loneliness and attention to social threat in young adults: Findings from an eye tracker study. *Personality and Individual Differences*, *63*, 16–23.
25. Prinstein, M. (2018). *The Popularity Illusion*, Kindle edition, location 2110. London: Ebury.
26. See the following for a summary of these perceptual effects, their implications for issues like anxiety and depression, and the potential treatment: Herz, N., Baror, S., and Bar, M. (2020). Overarching states of mind. *Trends in Cognitive Sciences*, *24*(3), 184–99; Kube, T., Schwarting, R., Rozenkrantz, L., Glombiewski, J.A., and Rief, W. (2020). Distorted cognitive processes in major depression: A predictive processing perspective. *Biological Psychiatry*, *87*(5), 388–98; Sussman, T.J., Jin, J., and Mohanty, A. (2016). Top-down and bottom-up factors in threat-related perception and attention in anxiety. *Biological Psychology*, *121*(Pt B), 160–72.
27. Shiban, Y., Fruth, M.B., Pauli, P., Kinateder, M., Reichenberger, J., and Mühlberger, A. (2016). Treatment effect on biases in size estimation in spider phobia. *Biological Psychology*, *121*(Pt B), 146–52.
28. Dennis, T.A., and O'Toole, L.J. (2014). Mental health on the go: Effects of a gamified attention-bias modification mobile application in trait-anxious adults. *Clinical Psychological Science*, *2*(5), 576–90; Mogg, K., and Bradley, B.P. (2016). Anxiety and attention to threat: Cognitive mechanisms and treatment with attention bias modification. *Behaviour Research and Therapy*, *87*, 76–108; Kress, L., and Aue, T. (2019). Learning to look at the bright side of life: Attention bias modification training enhances optimism bias. *Frontiers*

*in Human Neuroscience*, *13*, 222; Kuckertz, J.M., Schofield, C.A., Clerkin, E.M., Primack, J., Boettcher, H., Weisberg, R.B., . . . and Beard, C. (2019). Attentional bias modification for social anxiety disorder: What do patients think and why does it matter? *Behavioural and Cognitive Psychotherapy*, *47*(1), 16–38; Abado, E., Aue, T., and Okon-Singer, H. (2020). The missing pieces of the puzzle: A review on the interactive nature of a-priori expectancies and attention bias toward threat. *Brain Sciences*, *10*(10), 745; Jones, E.B., and Sharpe, L. (2017). Cognitive bias modification: A review of meta-analyses. *Journal of Affective Disorders*, *223*, 175–83; Gober, C. D., Lazarov, A., and Bar-Haim, Y. (2021). From cognitive targets to symptom reduction: Overview of attention and interpretation bias modification research. *Evidence-Based Mental Health*, *24*(1), 42–46.

29. See the following for a thorough description of gustatory expectation effects and their relation to predictive coding: Piqueras-Fiszman, B., and Spence, C. (2015). Sensory expectations based on product-extrinsic food cues: An interdisciplinary review of the empirical evidence and theoretical accounts. *Food Quality and Preference*, *40*, 165–79.
30. Spence, C., and Piqueras-Fiszman, B. (2014). *The Perfect Meal: The Multisensory Science of Food and Dining*. Chichester: John Wiley and Sons.
31. Lee, L., Frederick, S., and Ariely, D. (2006). Try it, you'll like it: The influence of expectation, consumption, and revelation on preferences for beer. *Psychological Science*, *17*(12), 1054–58.
32. Plassmann, H., O'Doherty, J., Shiv, B., and Rangel, A. (2008). Marketing actions can modulate neural representations of experienced pleasantness. *Proceedings of the National Academy of Sciences*, *105*(3), 1050–54.
33. Clark, A. (2016). *Surfing Uncertainty: Prediction, Action, and the Embodied Mind*, 55–56. Oxford: Oxford University Press.
34. Grabenhorst, F., Rolls, E.T., and Bilderbeck, A. (2007). How cognition modulates affective responses to taste and flavor: Top-down influences on the orbitofrontal and pregenual cingulate cortices. *Cerebral Cortex*, *18*(7), 1549–59.
35. Herz, R.S., and von Clef, J. (2001). The influence of verbal labeling on the perception of odors: Evidence for olfactory illusions? *Perception*, *30*(3), 381–91.
36. Fuller, T. (2013, December 3). A love letter to a smelly fruit. *New York Times*. https://www.nytimes.com/2013/12/08/travel/a-love-letter-to-a-smelly-fruit.html.
37. Amar, M., Ariely, D., Bar-Hillel, M., Carmon, Z., and Ofir, C. (2011). *Brand Names Act Like Marketing Placebos*. Available at http://www.ratio.huji.ac.il/sites/default/files/publications/dp566.pdf.
38. Langer, E., Djikic, M., Pirson, M., Madenci, A., and Donohue, R. (2010). Believing is seeing: Using mindlessness (mindfully) to improve visual acuity. *Psychological Science*, *21*(5), 661–66. See also Pirson, M., Ie, A., and Langer, E. (2012). Seeing what we know, knowing what we see: Challenging the limits of visual acuity. *Journal of Adult Development*, *19*(2), 59–65. Some may argue that the differences in visual acuity are merely "imagined." For an elegant experiment demonstrating that top-down processing can produce *objectively* sharper vision, see Lupyan, G. (2017). Objective effects of knowledge on visual perception. *Journal of Experimental Psychology: Human Perception and Performance*, *43*(4), 794.

## Chapter 2: A Pious Fraud

1. Blease, C., Annoni, M., and Hutchinson, P. (2018). Editors' introduction to special section on meaning response and the placebo effect. *Perspectives in Biology and Medicine*, *61*(3), 349–52. See also letter from Thomas Jefferson to Caspar Wistar, June 21, 1807. Available at: http://memory.loc.gov/service/mss/mtj/mtj1/038/038_0687_0692.pdf.

2. Raglin, J., Szabo, A., Lindheimer, J.B., and Beedie, C. (2020). Understanding placebo and nocebo effects in the context of sport: A psychological perspective. *European Journal of Sport Science*, *20*(3), 293–301; Aronson, J. (1999). Please, please me. *BMJ*, *318*(7185), 716; Kaptchuk, T.J. (1998). Powerful placebo: The dark side of the randomised controlled trial. *Lancet*, *351*(9117), 1722–25; De Craen, A.J., Kaptchuk, T.J., Tijssen, J.G., and Kleijnen, J. (1999). Placebos and placebo effects in medicine: Historical overview. *Journal of the Royal Society of Medicine*, *92*(10), 511–15.
3. Details of Beecher's wartime experiments, and his overall influence in medicine, can be found in the following: Beecher, H.K. (1946). Pain in men wounded in battle. *Annals of Surgery*, *123*(1), 96–105; Benedetti, F. (2016). Beecher as clinical investigator: Pain and the placebo effect. *Perspectives in Biology and Medicine*, *59*(1), 37–45; Gross, L. (2017). Putting placebos to the test. *PLoS Biology*, *15*(2), e2001998; Evans, D. (2003). *Placebo*. London: HarperCollins; Best, M., and Neuhauser, D. (2010). Henry K. Beecher: Pain, belief and truth at the bedside. The powerful placebo, ethical research and anaesthesia safety. *BMJ Quality and Safety*, *19*(5), 466–68.
4. Colloca, L. (2019). The placebo effect in pain therapies. *Annual Review of Pharmacology and Toxicology 59*, 191–211.
5. https://www.apdaparkinson.org/article/the-placebo-effect-in-clinical-trials-in-parkinsons-disease.
6. Lidstone, S.C., Schulzer, M., Dinelle, K., Mak, E., Sossi, V., Ruth, T.J., . . . and Stoessl, A.J. (2010). Effects of expectation on placebo-induced dopamine release in Parkinson disease. *Archives of General Psychiatry*, *67*(8), 857–65; Quattrone, A., Barbagallo, G., Cerasa, A., and Stoessl, A.J. (2018). Neurobiology of placebo effect in Parkinson's disease: What we have learned and where we are going. *Movement Disorders*, *33*(8), 1213–27.
7. Vits, S., Cesko, E., Benson, S., Rueckert, A., Hillen, U., Schadendorf, D., and Schedlowski, M. (2013). Cognitive factors mediate placebo responses in patients with house dust mite allergy. *PLoS One*, *8*(11), e79576. It's worth noting that various factors may influence the placebo responses here, including the patient's existing beliefs and the attitude of the physician. See Howe, L.C., Goyer, J.P., and Crum, A.J. (2017). Harnessing the placebo effect: Exploring the influence of physician characteristics on placebo response. *Health Psychology*, *36*(11), 1074–82; Leibowitz, K.A., Hardebeck, E.J., Goyer, J.P., and Crum, A.J. (2019). The role of patient beliefs in open-label placebo effects. *Health Psychology*, *38*(7), 613–22; Darragh, M., Chang, J.W., Booth, R.J., and Consedine, N.S. (2015). The placebo effect in inflammatory skin reactions: The influence of verbal suggestion on itch and weal size. *Journal of Psychosomatic Research*, *78*(5), 489–94; Pfaar, O., Agache, I., Bergmann, K.C., Bindslev-Jensen, C., Bousquet, J., Creticos, P.S., . . . and Frew, A.J. (2020). Placebo effects in allergen immunotherapy: An EAACI Task Force Position Paper. *Allergy*, *76*(3), 629–47.
8. Kemeny, M.E., Rosenwasser, L.J., Panettieri, R.A., Rose, R.M., Berg-Smith, S.M., and Kline, J.N. (2007). Placebo response in asthma: A robust and objective phenomenon. *Journal of Allergy and Clinical Immunology*, *119*(6), 1375–81. Placebos seem to have very large effects on patients' subjective distress, but the differences can also be noted in objective measures of their breathing. See Luc, F., Prieur, E., Whitmore, G.A., Gibson, P.G., Vandemheen, K.L., and Aaron, S.D. (2019). Placebo effects in clinical trials evaluating patients with uncontrolled persistent asthma. *Annals of the American Thoracic Society*, *16*(9), 1124–30.
9. Al-Lamee, R., Thompson, D., Dehbi, H.M., Sen, S., Tang, K., Davies, J., . . . and Nijjer, S.S. (2018). Percutaneous coronary intervention in stable angina (ORBITA): A double-blind, randomised controlled trial. *Lancet*, *391*(10115), 31–40.
10. Horwitz, R.I., Viscoli, C.M., Donaldson, R.M., Murray, C.J., Ransohoff, D.F., Berkman, L., . . . and Sindelar, J. (1990). Treatment adherence and risk of death after a myocardial

infarction. *Lancet*, *336*(8714), 542–45; for a discussion, see Brown, W.A. (1998). Harnessing the placebo effect. *Hospital Practice*, *33*(7), 107–16.

11. See, for instance, Simpson, S.H., Eurich, D.T., Majumdar, S.R., Padwal, R.S., Tsuyuki, R.T., Varney, J., and Johnson, J.A. (2006). A meta-analysis of the association between adherence to drug therapy and mortality. *BMJ*, *333*(7557), 15; Pressman, A., Avins, A.L., Neuhaus, J., Ackerson, L., and Rudd, P. (2012). Adherence to placebo and mortality in the Beta Blocker Evaluation of Survival Trial (BEST). *Contemporary Clinical Trials*, *33*(3), 492–98.
12. This argument has been proposed by numerous scientists. See Moerman, D.E. (2002). *Meaning, Medicine, and the "Placebo Effect,"* 116–21. Cambridge: Cambridge University Press; Chewning, B. (2006). The healthy adherer and the placebo effect. *BMJ*, *333*(7557), 18; Wilson, I.B. (2010). Adherence, placebo effects, and mortality. *Journal of General Internal Medicine*, *25*(12), 1270–72; Yue, Z., Cai, C., Ai-Fang, Y., Feng-Min, T., Li, C., and Bin, W. (2014). The effect of placebo adherence on reducing cardiovascular mortality: A meta-analysis. *Clinical Research in Cardiology*, *103*(3), 229–35.
13. The preceding three paragraphs synthesize various explanations for the placebo effect, including Petrie, K.J., and Rief, W. (2019). Psychobiological mechanisms of placebo and nocebo effects: Pathways to improve treatments and reduce side effects. *Annual Review of Psychology*, *70*, 599–625; Colloca, L., and Barsky, A.J. (2020). Placebo and nocebo effects. *New England Journal of Medicine*, *382*(6), 554–61; Colagiuri, B., Schenk, L.A., Kessler, M.D., Dorsey, S.G., and Colloca, L. (2015). The placebo effect: From concepts to genes. *Neuroscience*, *307*, 171–90; Ongaro, G., and Kaptchuk, T.J. (2019). Symptom perception, placebo effects, and the Bayesian brain. *Pain*, *160*(1), 1; Koban, L., Jepma, M., López-Solà, M., and Wager, T.D. (2019). Different brain networks mediate the effects of social and conditioned expectations on pain. *Nature Communications*, *10*(1), 1–13; Miller, F.G., Colloca, L., and Kaptchuk, T.J. (2009). The placebo effect: Illness and interpersonal healing. *Perspectives in Biology and Medicine*, *52*(4), 518; Trimmer, P.C., Marshall, J.A., Fromhage, L., McNamara, J.M., and Houston, A.I. (2013). Understanding the placebo effect from an evolutionary perspective. *Evolution and Human Behavior*, *34*(1), 8–15; Meissner, K. (2011). The placebo effect and the autonomic nervous system: Evidence for an intimate relationship. *Philosophical Transactions of the Royal Society B: Biological Sciences*, *366*(1572), 1808–17.
14. Crum, A.J., Phillips, D.J., Goyer, J.P., Akinola, M., and Higgins, E.T. (2016). Transforming water: Social influence moderates psychological, physiological, and functional response to a placebo product. *PLoS One*, *11*(11), e0167121. See also https://sparq.stanford.edu/director-crum-publishes-intriguing-study-placebo-effects.
15. Ho, J.T., Krummenacher, P., Lesur, M.R., and Lenggenhager, B. (2020). Real bodies not required? Placebo analgesia and pain perception in immersive virtual and augmented reality. *bioRxiv*. https://www.biorxiv.org/content/10.1101/2020.12.18.423276v1.abstract.
16. Buckalew, L.W., and Ross, S. (1981). Relationship of perceptual characteristics to efficacy of placebos. *Psychological Reports*, *49*(3), 955–61.
17. Faasse, K., and Martin, L.R. (2018). The power of labeling in nocebo effects. *International Review of Neurobiology*, *139*, 379–406.
18. Faasse, K., Martin, L.R., Grey, A., Gamble, G., and Petrie, K.J. (2016). Impact of brand or generic labeling on medication effectiveness and side effects. *Health Psychology*, *35*(2), 187.
19. Walach, H., and Jonas, W.B. (2004). Placebo research: The evidence base for harnessing self-healing capacities. *Journal of Alternative and Complementary Medicine*, *10* (Supplement 1), S-103.
20. Howe, L.C., Goyer, J.P., and Crum, A.J. (2017). Harnessing the placebo effect: Exploring the influence of physician characteristics on placebo response. *Health Psychology*, *36*(11), 1074.

21. Howick, J., Bishop, F.L., Heneghan, C., Wolstenholme, J., Stevens, S., Hobbs, F.R., and Lewith, G. (2013). Placebo use in the United Kingdom: Results from a national survey of primary care practitioners. *PLoS One*, *8*(3), e58247.
22. Silberman, S. (2009). Placebos are getting more effective. Drug makers are desperate to know why. *Wired Magazine*, *17*, 1–8.
23. Walsh, B.T., Seidman, S.N., Sysko, R., and Gould, M. (2002). Placebo response in studies of major depression: Variable, substantial, and growing. *JAMA*, *287*(14), 1840–47; Dunlop, B. W., Thase, M.E., Wun, C.C., Fayyad, R., Guico-Pabia, C.J., Musgnung, J., and Ninan, P.T. (2012). A meta-analysis of factors impacting detection of antidepressant efficacy in clinical trials: The importance of academic sites. *Neuropsychopharmacology*, *37*(13), 2830–36.
24. Tuttle, A.H., Tohyama, S., Ramsay, T., Kimmelman, J., Schweinhardt, P., Bennett, G.J., and Mogil, J.S. (2015). Increasing placebo responses over time in US clinical trials of neuropathic pain. *Pain*, *156*(12), 2616–26. For a breakdown of the statistics, see Marchant, J. (2015). Strong placebo response thwarts painkiller trials. *Nature News*. doi: 10.1038/nature.2015.18511.
25. Bennett, G.J. (2018). Does the word "placebo" evoke a placebo response? *Pain*, *159*(10), 1928–31.
26. Beecher, H.K. (1955). The powerful placebo. *Journal of the American Medical Association*, *159*(17), 1602–6. (The emphasis, within the quote, is my own.)
27. For evidence that an explanation can heighten the effects of open-label placebos, see Locher, C., Nascimento, A.F., Kirsch, I., Kossowsky, J., Meyer, A., and Gaab, J. (2017). Is the rationale more important than deception? A randomized controlled trial of open-label placebo analgesia. *Pain*, *158*(12), 2320–28; Wei, H., Zhou, L., Zhang, H., Chen, J., Lu, X., and Hu, L. (2018). The influence of expectation on nondeceptive placebo and nocebo effects. *Pain Research and Management*. doi: 10.1155/2018/8459429.
28. Carvalho, C., Caetano, J.M., Cunha, L., Rebouta, P., Kaptchuk, T.J., and Kirsch, I. (2016). Open-label placebo treatment in chronic low back pain: A randomized controlled trial. *Pain*, *157*(12), 2766.
29. Carvalho, C., Pais, M., Cunha, L., Rebouta, P., Kaptchuk, T.J., and Kirsch, I. (2020). Open-label placebo for chronic low back pain: A 5-year follow-up. *Pain*, *162*(5), 1521–27.
30. Kaptchuk, T.J., and Miller, F.G. (2018). Open label placebo: Can honestly prescribed placebos evoke meaningful therapeutic benefits? *BMJ*, *363*, k3889. doi: 10.1136/bmj.k3889.
31. Schaefer, M., Sahin, T., and Berstecher, B. (2018). Why do open-label placebos work? A randomized controlled trial of an open-label placebo induction with and without extended information about the placebo effect in allergic rhinitis. *PLoS One*, *13*(3), e0192758.
32. Bernstein, M.H., Magill, M., Beaudoin, F.L., Becker, S.J., and Rich, J.D. (2018). Harnessing the placebo effect: A promising method for curbing the opioid crisis? *Addiction*, *113*(11), 2144–45.
33. CDC, Opioid data analysis and resources, https://www.cdc.gov/drugoverdose/data/analysis.html.
34. Morales-Quezada, L., Mesia-Toledo, I., Estudillo-Guerra, A., O'Connor, K.C., Schneider, J.C., Sohn, D.J., . . . and Zafonte, R. (2020). Conditioning open-label placebo: A pilot pharmacobehavioral approach for opioid dose reduction and pain control. *Pain Reports*, *5*(4). See also Flowers, K.M., Patton, M.E., Hruschak, V.J., Fields, K.G., Schwartz, E., Zeballos, J., . . . and Schreiber, K.L. (2021). Conditioned open-label placebo for opioid reduction after spine surgery: A randomized controlled trial. *Pain*, *162*(6), 1828–39.
35. Laferton, J.A., Mora, M.S., Auer, C.J., Moosdorf, R., and Rief, W. (2013). Enhancing the efficacy of heart surgery by optimizing patients' preoperative expectations: Study protocol

of a randomized controlled trial. *American Heart Journal*, *165*(1), 1–7. See the following for a more elaborate description of the theory behind these kinds of interventions: Doering, B.K., Glombiewski, J.A., and Rief, W. (2018). Expectation-focused psychotherapy to improve clinical outcomes. *International Review of Neurobiology*, *138*, 257–70.

36. Auer, C.J., Laferton, J.A., Shedden-Mora, M. C., Salzmann, S., Moosdorf, R., and Rief, W. (2017). Optimizing preoperative expectations leads to a shorter length of hospital stay in CABG patients: Further results of the randomized controlled PSY-HEART trial. *Journal of Psychosomatic Research*, *97*, 82–89.
37. Rief, W., Shedden-Mora, M.C., Laferton, J.A., Auer, C., Petrie, K.J., Salzmann, S., . . . and Moosdorf, R. (2017). Preoperative optimization of patient expectations improves long-term outcome in heart surgery patients: Results of the randomized controlled PSY-HEART trial. *BMC Medicine*, *15*(1), 1–13.
38. For further evidence of the potential for people's expectations to shape the success of surgical procedures, see Auer, C.J., Glombiewski, J.A., Doering, B.K., Winkler, A., Laferton, J.A., Broadbent, E., and Rief, W. (2016). Patients' expectations predict surgery outcomes: A meta-analysis. *International Journal of Behavioral Medicine*, *23*(1), 49–62; Kube, T., Glombiewski, J.A., and Rief, W. (2018). Using different expectation mechanisms to optimize treatment of patients with medical conditions: A systematic review. *Psychosomatic Medicine*, *80*(6), 535–43; Van Der Meij, E., Anema, J.R., Leclercq, W.K., Bongers, M.Y., Consten, E.C., Koops, S.E.S., . . . and Huirne, J.A. (2018). Personalised perioperative care by e-health after intermediate-grade abdominal surgery: A multicentre, single-blind, randomised, placebo-controlled trial. *Lancet*, *392*(10141), 51–59; Laferton, J.A., Oeltjen, L., Neubauer, K., Ebert, D.D., and Munder, T. (2020). The effects of patients' expectations on surgery outcome in total hip and knee arthroplasty: A prognostic factor meta-analysis. *Health Psychology Review*. doi: 10.1080/17437199.2020.1854051.
39. Akroyd, A., Gunn, K.N., Rankin, S., Douglas, M., Kleinstäuber, M., Rief, W., and Petrie, K.J. (2020). Optimizing patient expectations to improve therapeutic response to medical treatment: A randomized controlled trial of iron infusion therapy. *British Journal of Health Psychology*, *25*(3), 639–51.
40. Leibowitz, K.A., Hardebeck, E.J., Goyer, J.P., and Crum, A.J. (2018). Physician assurance reduces patient symptoms in US adults: An experimental study. *Journal of General Internal Medicine*, *33*(12), 2051–52.
41. Rakel, D., Barrett, B., Zhang, Z., Hoeft, T., Chewning, B., Marchand, L., and Scheder, J. (2011). Perception of empathy in the therapeutic encounter: Effects on the common cold. *Patient Education and Counseling*, *85*(3), 390–97.

## Chapter 3: Do No Harm

1. Rose, R. (1956). *Living Magic: The Realities Underlying the Psychical Practices and Beliefs of Australian Aborigines*, 28–47. New York: Rand McNally.
2. See also Cannon, W.B. (1942). "Voodoo" death. *American Anthropologist*, *44*(2), 169–81; Benson, H. (1997). The nocebo effect: History and physiology. *Preventive Medicine*, *26*(5), 612–15; Byard, R. (1988). Traditional medicine of aboriginal Australia. *CMAJ: Canadian Medical Association Journal*, *139*(8), 792. For a discussion of alternative explanations of these deaths: Lester, D. (2009). Voodoo death. *OMEGA: Journal of Death and Dying*, *59*(1), 1–18.
3. For a summary of medical theories of voodoo death, see Samuels, M.A. (2007). "Voodoo" death revisited: The modern lessons of neurocardiology. *Cleveland Clinic Journal of Medicine*, *74*(Suppl 1), S8–S16; Morse, D.R., Martin, J., and Moshonov, J. (1991). Psychosomatically induced death relative to stress, hypnosis, mind control, and voodoo: Review and possible mechanisms. *Stress Medicine*, *7*(4), 213–32.

4. Meador, C.K. (1992). Hex death: Voodoo magic or persuasion? *Southern Medical Journal, 85*(3), 244–47.
5. Milton, G.W. (1973). Self-willed death or the bone-pointing syndrome. *Lancet, 301*(7817), 1435–36. For many similar accounts, see Benson, H. (1997). The nocebo effect: History and physiology. *Preventive Medicine, 26*(5), 612–15.
6. The potential link between the nocebo effect and voodoo death is very widely recognized. See, for example, Edwards, I.R., Graedon, J., and Graedon, T. (2010). Placebo harm. *Drug Safety, 33*(6), 439–41; Benedetti, F. (2013). Placebo and the new physiology of the doctor–patient relationship. *Physiological Reviews, 93*(3), 1207–46; Cheyne, J.A., and Pennycook, G. (2013). Sleep paralysis postepisode distress: Modeling potential effects of episode characteristics, general psychological distress, beliefs, and cognitive style. *Clinical Psychological Science, 1*(2), 135–48.
7. Mackenzie, J.N. (1886). The production of the so-called "rose cold" by means of an artificial rose, with remarks and historical notes. *American Journal of the Medical Sciences, 91*(181), 45. While this is based on a single anecdote, modern research shows that the mere expectation of a hay fever attack can indeed bring about symptoms in sufferers: Besedovsky, L., Benischke, M., Fischer, J., Yazdi, A.S., and Born, J. (2020). Human sleep consolidates allergic responses conditioned to the environmental context of an allergen exposure. *Proceedings of the National Academy of Sciences, 117*(20), 10983–88. See also Jewett, D.L., Fein, G., and Greenberg, M.H. (1990). A double-blind study of symptom provocation to determine food sensitivity. *New England Journal of Medicine, 323*(7), 429–33.
8. Beecher, H.K. (1955). The powerful placebo. *Journal of the American Medical Association, 159*(17), 1602–6.
9. Howick, J., Webster, R., Kirby, N., and Hood, K. (2018). Rapid overview of systematic reviews of nocebo effects reported by patients taking placebos in clinical trials. *Trials, 19*(1), 1–8; see also Mahr, A., Golmard, C., Pham, E., Iordache, L., Deville, L., and Faure, P. (2017). Types, frequencies, and burden of nonspecific adverse events of drugs: Analysis of randomized placebo-controlled clinical trials. *Pharmacoepidemiology and Drug Safety, 26*(7), 731–41.
10. https://www.nhs.uk/medicines/finasteride.
11. Mondaini, N., Gontero, P., Giubilei, G., Lombardi, G., Cai, T., Gavazzi, A., and Bartoletti, R. (2007). Finasteride 5 mg and sexual side effects: How many of these are related to a nocebo phenomenon? *Journal of Sexual Medicine, 4*(6), 1708–12.
12. Myers, M.G., Cairns, J.A., and Singer, J. (1987). The consent form as a possible cause of side effects. *Clinical Pharmacology and Therapeutics, 42*(3), 250–53.
13. Varelmann, D., Pancaro, C., Cappiello, E.C., and Camann, W.R. (2010). Nocebo-induced hyperalgesia during local anesthetic injection. *Anesthesia and Analgesia, 110*(3), 868–70.
14. Tinnermann, A., Geuter, S., Sprenger, C., Finsterbusch, J., and Büchel, C. (2017). Interactions between brain and spinal cord mediate value effects in nocebo hyperalgesia. *Science, 358*(6359), 105–8.
15. Aslaksen, P.M., Zwarg, M.L., Eilertsen, H.-I. H., Gorecka, M.M., and Bjørkedal, E. (2015). Opposite effects of the same drug. *Pain, 156*(1), 39–46; Flaten, M.A., Simonsen, T., and Olsen, H. (1999). Drug-related information generates placebo and nocebo responses that modify the drug response. *Psychosomatic Medicine, 61*(2), 250–55.
16. Scott, D.J., Stohler, C.S., Egnatuk, C.M., Wang, H., Koeppe, R.A., and Zubieta, J.K. (2008). Placebo and nocebo effects are defined by opposite opioid and dopaminergic responses. *Archives of General Psychiatry, 65*(2), 220–31.
17. Enck, P., Benedetti, F., and Schedlowski, M. (2008). New insights into the placebo and nocebo responses. *Neuron, 59*(2), 195–206.

18. Planès, S., Villier, C., and Mallaret, M. (2016). The nocebo effect of drugs. *Pharmacology Research and Perspectives*, *4*(2), e00208; Liccardi, G., Senna, G., Russo, M., Bonadonna, P., Crivellaro, M., Dama, A., . . . and Passalacqua, G. (2004). Evaluation of the nocebo effect during oral challenge in patients with adverse drug reactions. *Journal of Investigational Allergology and Clinical Immunology 14*(2), 104–7.
19. Faasse, K., Cundy, T., Gamble, G., and Petrie, K.J. (2013). The effect of an apparent change to a branded or generic medication on drug effectiveness and side effects. *Psychosomatic Medicine*, *75*(1), 90–96.
20. Faasse, K., Cundy, T., and Petrie, K.J. (2009). Thyroxine: Anatomy of a health scare. *BMJ*, *339*, b5613. doi: 10.1136/bmj.b5613. See also Faasse, K., Cundy, T., Gamble, G., and Petrie, K.J. (2013). The effect of an apparent change to a branded or generic medication on drug effectiveness and side effects. *Psychosomatic Medicine*, *75*(1), 90–96; MacKrill, K., and Petrie, K.J. (2018). What is associated with increased side effects and lower perceived efficacy following switching to a generic medicine? A New Zealand cross-sectional patient survey. *BMJ Open*, *8*(10), e023667. For a full analysis, see Faasse, K., and Martin, L.R. (2018). The power of labeling in nocebo effects. *International Review of Neurobiology*, *139*, 379–406.
21. Blasini, M., Corsi, N., Klinger, R., and Colloca, L. (2017). Nocebo and pain: An overview of the psychoneurobiological mechanisms. *Pain Reports*, *2*(2).
22. Sciama, Y. (2017, September 27). France brings back a phased-out drug after patients rebel against its replacement. *Science*. https://www.sciencemag.org/news/2017/09/france-brings-back-phased-out-drug-after-patients-rebel-against-its-replacement.
23. Rippon, G. (2019). *The Gendered Brain*, 29. London: Bodley Head; Ruble, D.N. (1977). Premenstrual symptoms: A reinterpretation. *Science*, *197*(4300), 291–92.
24. Horing, B., Weimer, K., Schrade, D., Muth, E.R., Scisco, J.L., Enck, P., and Klosterhalfen, S. (2013). Reduction of motion sickness with an enhanced placebo instruction: An experimental study with healthy participants. *Psychosomatic Medicine*, *75*(5), 497–504; Eden, D., and Zuk, Y. (1995). Seasickness as a self-fulfilling prophecy: Raising self-efficacy to boost performance at sea. *Journal of Applied Psychology*, *80*(5), 628.
25. Ferrari, R., Obelieniene, D., Darlington, P., Gervais, R., and Green, P. (2002). Laypersons' expectation of the sequelae of whiplash injury: A cross-cultural comparative study between Canada and Lithuania. *Medical Science Monitor*, *8*(11), CR728–CR734; Buchbinder, R., and Jolley, D. (2005). Effects of a media campaign on back beliefs is sustained three years after its cessation. *Spine*, *30*(11), 1323–30; Polich, G., Iaccarino, M.A., Kaptchuk, T.J., Morales-Quezada, L., and Zafonte, R. (2020). Nocebo effects in concussion: Is all that is told beneficial? *American Journal of Physical Medicine and Rehabilitation*, *99*(1), 71–80.
26. Whittaker, R., Kemp, S., and House, A. (2007). Illness perceptions and outcome in mild head injury: A longitudinal study. *Journal of Neurology, Neurosurgery and Psychiatry*, *78*(6), 644–46. See also Hou, R., Moss-Morris, R., Peveler, R., Mogg, K., Bradley, B. P., and Belli, A. (2012). When a minor head injury results in enduring symptoms: A prospective investigation of risk factors for postconcussional syndrome after mild traumatic brain injury. *Journal of Neurology, Neurosurgery and Psychiatry*, *83*(2), 217–23.
27. Polich, G., Iaccarino, M.A., Kaptchuk, T.J., Morales-Quezada, L., and Zafonte, R. (2020). Nocebo effects in concussion: Is all that is told beneficial? *American Journal of Physical Medicine and Rehabilitation*, *99*(1), 71–80.
28. Reeves, R.R., Ladner, M.E., Hart, R.H., and Burke, R.S. (2007). Nocebo effects with antidepressant clinical drug trial placebos. *General Hospital Psychiatry*, *29*(3), 275–77.
29. Usichenko, T.I., Hacker, H., and Hesse, T. (2016). Nocebo effect of informed consent: Circulatory collapse before elective caesarean section. *International Journal of Obstetric Anesthesia*, *27*, 95–96.

30. Samuels, M.A. (2007). Voodoo death revisited: The modern lessons of neurocardiology. *Cleveland Clinic Journal of Medicine*, *74* (Suppl 1), S8–S16. See also Amanzio, M., Howick, J., Bartoli, M., Cipriani, G.E., and Kong, J. (2020). How do nocebo phenomena provide a theoretical framework for the COVID–19 pandemic? *Frontiers in Psychology*, *11*, 589884.
31. Eaker, E.D., Pinsky, J., and Castelli, W.P. (1992). Myocardial infarction and coronary death among women: Psychosocial predictors from a 20-year follow-up of women in the Framingham Study. *American Journal of Epidemiology*, *135*(8), 854–64. See also Olshansky, B. (2007). Placebo and nocebo in cardiovascular health: Implications for healthcare, research, and the doctor-patient relationship. *Journal of the American College of Cardiology*, *49*(4), 415–21.
32. Barefoot, J.C., Brummett, B.H., Williams, R.B., Siegler, I.C., Helms, M.J., Boyle, S.H., . . . and Mark, D.B. (2011). Recovery expectations and long-term prognosis of patients with coronary heart disease. *Archives of Internal Medicine*, *171*(10), 929–35.
33. Carey, I.M., Shah, S.M., DeWilde, S., Harris, T., Victor, C.R., and Cook, D.G. (2014). Increased risk of acute cardiovascular events after partner bereavement: A matched cohort study. *JAMA Internal Medicine*, *174*(4), 598–605.
34. Shimizu, M., and Pelham, B.W. (2008). Postponing a date with the grim reaper: Ceremonial events and mortality. *Basic and Applied Social Psychology*, *30*(1), 36–45; Wilches-Gutiérrez, J.L., Arenas-Monreal, L., Paulo-Maya, A., Peláez-Ballestas, I., and Idrovo, A.J. (2012). A "beautiful death": Mortality, death, and holidays in a Mexican municipality. *Social Science and Medicine*, *74*(5), 775–82; Ajdacic-Gross, V., Knöpfli, D., Landolt, K., Gostynski, M., Engelter, S.T., Lyrer, P.A., . . . and Rössler, W. (2012). Death has a preference for birthdays: An analysis of death time series. *Annals of Epidemiology*, *22*(8), 603–6; Kelly, G.E., and Kelleher, C.C. (2018). Happy birthday? An observational study. *Journal of Epidemioliogy and Community Health*, *72*(12), 1168–72. See also Phillips, D.P., and Feldman, K.A. (1973). A dip in deaths before ceremonial occasions: Some new relationships between social integration and mortality. *American Sociological Review*, *38*(6), 678–96; Byers, B., Zeller, R.A., and Byers, P.Y. (1991). Birthdate and mortality: An evaluation of the death-dip/death-rise phenomenon. *Sociological Focus*, *24*(1), 13–28; Phillips, D.P., Van Voorhees, C.A., and Ruth, T.E. (1992). The birthday: Lifeline or deadline? *Psychosomatic Medicine*, *54*(5), 532–42.
35. National Constitution Center. (2020). Three presidents die on July 4th: Just a coincidence? https://constitutioncenter.org/blog/three-presidents-die-on-july-4th-just-a-coincidence.
36. See the following for a broad discussion of all these phenomena: Ray, O. (2004). How the mind hurts and heals the body. *American Psychologist*, *59*(1), 29.
37. Pan, Y., Kinitz, T., Stapic, M., and Nestoriuc, Y. (2019). Minimizing drug adverse events by informing about the nocebo effect: An experimental study. *Frontiers in Psychiatry*, *10*, 504.
38. Howick, J. (2020). Unethical informed consent caused by overlooking poorly measured nocebo effects. *Journal of Medical Ethics*. doi: 10.1136/medethics-2019–105903; see also Colloca, L. (2017). Tell me the truth and I will not be harmed: Informed consents and nocebo effects. *American Journal of Bioethics*, *17*(6), 46–48.
39. Faasse, K., Huynh, A., Pearson, S., Geers, A.L., Helfer, S.G., and Colagiuri, B. (2019). The influence of side effect information framing on nocebo effects. *Annals of Behavioral Medicine*, *53*(7), 621–29.
40. James, L.K., and Till, S.J. (2016). Potential mechanisms for IgG4 inhibition of immediate hypersensitivity reactions. *Current Allergy and Asthma Reports*, *16*(3), 1–7; Couzin-Frankel, J. (2018, October 18). A revolutionary treatment for allergies to peanuts and other foods is going mainstream. *Science*. https://www.sciencemag.org/news/2018/10

/revolutionary-treatment-allergies-peanuts-and-other-foods-going-mainstream-do-benefits.

41. Howe, L.C., Leibowitz, K.A., Perry, M.A., Bitler, J.M., Block, W., Kaptchuk, T.J., . . . and Crum, A.J. (2019). Changing patient mindsets about non-life-threatening symptoms during oral immunotherapy: A randomized clinical trial. *Journal of Allergy and Clinical Immunology: In Practice*, *7*(5), 1550–59; Positive mindset about side effects of peanut-allergy treatment improves outcomes. (2019, February 5), Stanford Medicine News Center. https://med.stanford.edu/news/all-news/2019/02/positive-mindset-about-side-effects-of-peanut-allergy-treatment.html. See the following for a broader discussion of these mindset effects and their therapeutic potential: Leibowitz, K.A., Howe, L.C., and Crum, A.J. (2021). Changing mindsets about side effects. *BMJ Open*, *11*(2), e040134.
42. For evidence of pain catastrophizing's effects on opioid signaling, see King, C.D., Goodin, B., Kindler, L.L., Caudle, R.M., Edwards, R.R., Gravenstein, N., . . . and Fillingim, R.B. (2013). Reduction of conditioned pain modulation in humans by naltrexone: An exploratory study of the effects of pain catastrophizing. *Journal of Behavioral Medicine*, *36*(3), 315–27; Vögtle, E., Barke, A., and Kröner-Herwig, B. (2013). Nocebo hyperalgesia induced by social observational learning. *Pain*, *154*(8), 1427–33.
43. Granot, M., and Ferber, S.G. (2005). The roles of pain catastrophizing and anxiety in the prediction of postoperative pain intensity: A prospective study. *Clinical Journal of Pain*, *21*(5), 439–45; Witvrouw, E., Pattyn, E., Almqvist, K.F., Crombez, G., Accoe, C., Cambier, D., and Verdonk, R. (2009). Catastrophic thinking about pain as a predictor of length of hospital stay after total knee arthroplasty: A prospective study. *Knee Surgery, Sports Traumatology, Arthroscopy*, *17*(10), 1189–94.
44. Drahovzal, D.N., Stewart, S.H., and Sullivan, M.J. (2006). Tendency to catastrophize somatic sensations: Pain catastrophizing and anxiety sensitivity in predicting headache. *Cognitive Behaviour Therapy*, *35*(4), 226–35; Mortazavi Nasiri, F.S., Pakdaman, S., Dehghani, M., and Togha, M. (2017). The relationship between pain catastrophizing and headache-related disability: The mediating role of pain intensity. *Japanese Psychological Research*, *59*(4), 266–74; Martinez-Calderon, J., Jensen, M.P., Morales-Asencio, J.M., and Luque-Suarez, A. (2019). Pain catastrophizing and function in individuals with chronic musculoskeletal pain. *Clinical Journal of Pain*, *35*(3), 279–93.
45. Darnall, B.D., and Colloca, L. (2018). Optimizing placebo and minimizing nocebo to reduce pain, catastrophizing, and opioid use: A review of the science and an evidence-informed clinical toolkit. *International Review of Neurobiology*, *139*, 129–57.
46. Ibid.
47. Seng, E.K. (2018). Using cognitive behavioral therapy techniques to treat migraine. *Journal of Health Service Psychology*, *44*(2), 68–73.
48. Ehde, D.M., and Jensen, M.P. (2004). Feasibility of a cognitive restructuring intervention for treatment of chronic pain in persons with disabilities. *Rehabilitation Psychology*, *49*(3), 254.
49. Lumley, M.A., and Schubiner, H. (2019). Psychological therapy for centralized pain: An integrative assessment and treatment model. *Psychosomatic Medicine*, *81*(2), 114–24.
50. Ibid. Similar results can be found for people with auto-immune disorders: Karademas, E.C., Dimitraki, G., Papastefanakis, E., Ktistaki, G., Repa, A., Gergianaki, I., . . . and Simos, P. (2018). Emotion regulation contributes to the well-being of patients with autoimmune diseases through illness-related emotions: A prospective study. *Journal of Health Psychology*, *25*(13–14), 2096–2105; Nahman-Averbuch, H., Schneider, V.J., Chamberlin, L.A., Van Diest, A.M.K., Peugh, J.L., Lee, G.R., . . . and King, C.D. (2021). Identification

of neural and psychophysical predictors of headache reduction after cognitive behavioral therapy in adolescents with migraine. *Pain, 162 (2)*, 372–81.

51. Adamczyk, A.K., Ligeza, T.S., and Wyczesany, M. (2020). The dynamics of pain reappraisal: The joint contribution of cognitive change and mental load. *Cognitive, Affective, and Behavioral Neuroscience*, *20*(2), 276–93.
52. De Peuter, S., Lemaigre, V., Van Diest, I., and Van den Bergh, O. (2008). Illness-specific catastrophic thinking and over-perception in asthma. *Health Psychology*, *27*(1), 93.
53. Brown, R.L., Shahane, A. D., Chen, M.A., and Fagundes, C.P. (2020). Cognitive reappraisal and nasal cytokine production following experimental rhinovirus infection. *Brain, Behavior, and Immunity-Health*, *1*, 100012.
54. Dekker, R.L., Moser, D.K., Peden, A.R., and Lennie, T.A. (2012). Cognitive therapy improves three-month outcomes in hospitalized patients with heart failure. *Journal of Cardiac Failure*, *18*(1), 10–20. See the following for the proposed physiological and behavioral mechanisms: Celano, C.M., Villegas, A.C., Albanese, A.M., Gaggin, H.K., and Huffman, J.C. (2018). Depression and anxiety in heart failure: A review. *Harvard Review of Psychiatry*, *26*(4), 175.

## Chapter 4: The Origins of Mass Hysteria

1. Escola encerra devido a alergis. (2006, May 18). *CM.* https://www.cmjornal.pt/portugal/detalhe/escola-encerra-devido-a-alergias; Televírus volta a atacar. (2006, May 18). *CM.* https://www.cmjornal.pt/portugal/detalhe/televirus-volta-a-atacar.
2. Bartholomew, R.E., Wessely, S., and Rubin, G.J. (2012). Mass psychogenic illness and the social network: Is it changing the pattern of outbreaks? *Journal of the Royal Society of Medicine*, *105*(12), 509–12.
3. Kilner, J.M., Friston, K.J., and Frith, C.D. (2007). Predictive coding: An account of the mirror neuron system. *Cognitive Processing*, *8*(3), 159–66.
4. See Di Pellegrino, G., Fadiga, L., Fogassi, L., Gallese, V., and Rizzolatti, G. (1992). Understanding motor events: A neurophysiological study. *Experimental Brain Research*, *91*(1), 176–80; Lametti, D. (2009, June 9). Mirroring behavior. *Scientific American*.
5. Bentivoglio, L. (2012, August 27). Rizzolatti: "Ecco perchè i sentimenti sono contagiosi." *La Repubblica*. https://parma.repubblica.it/cronaca/2012/08/27/news/rizzolatti_ecco_perch_i_sentimenti_sono_contagiosi-41547512.
6. Bastiaansen, J.A., Thioux, M., and Keysers, C. (2009). Evidence for mirror systems in emotions. *Philosophical Transactions of the Royal Society B: Biological Sciences*, *364*(1528), 2391–404.
7. Much of the research discussed in this section is covered in the following review paper: Hatfield, E., Carpenter, M., and Rapson, R.L. (2014). Emotional contagion as a precursor to collective emotions. In C. von Scheve and M. Salmela (Eds.). *Collective Emotions: Perspectives from Psychology, Philosophy, and Sociology*, 108–22. Oxford: Oxford University Press. For additional details, see Laird, J.D., Alibozak, T., Davainis, D., Deignan, K., Fontanella, K., Hong, J., . . . and Pacheco, C. (1994). Individual differences in the effects of spontaneous mimicry on emotional contagion. *Motivation and Emotion*, *18*(3), 231–47; Carsten, T., Desmet, C., Krebs, R.M., and Brass, M. (2018). Pupillary contagion is independent of the emotional expression of the face. *Emotion*, *19*(8), 1343–52.
8. Likowski, K.U., Mühlberger, A., Gerdes, A., Wieser, M.J., Pauli, P., and Weyers, P. (2012). Facial mimicry and the mirror neuron system: Simultaneous acquisition of facial electromyography and functional magnetic resonance imaging. *Frontiers in Human Neuroscience*, *6*, 214.
9. Neal, D.T., and Chartrand, T.L. (2011). Embodied emotion perception: Amplifying and dampening facial feedback modulates emotion perception accuracy. *Social Psychological and*

*Personality Science*, *2*(6), 673–78. For a recent replication, see Borgomaneri, S., Bolloni, C., Sessa, P., and Avenanti, A. (2020). Blocking facial mimicry affects recognition of facial and body expressions. *PLoS One*, *15*(2), e0229364; See also the following meta-analysis, which confirms the subtle effect of facial feedback on participants' emotions: Coles, N. A., Larsen, J. T., and Lench, H.C. (2019). A meta-analysis of the facial feedback literature: Effects of facial feedback on emotional experience are small and variable. *Psychological Bulletin*, *145*(6), 610.

10. Havas, D.A., Glenberg, A.M., and Rinck, M. (2007). Emotion simulation during language comprehension. *Psychonomic Bulletin and Review*, *14*(3), 436–41; Foroni, F., and Semin, G.R. (2009). Language that puts you in touch with your bodily feelings: The multimodal responsiveness of affective expressions. *Psychological Science*, *20*(8), 974–80.
11. Rizzolatti, G., Fogassi, L., and Gallese, V. (2006). Mirrors in the mind. *Scientific American*, *295*(5), 54–61.
12. Christakis, N.A., and Fowler, J.H. (2009). *Connected: The Surprising Power of Our Social Networks and How They Shape Our Lives*, 50–52. New York: Little, Brown Spark.
13. Faasse, K., and Petrie, K.J. (2016). From me to you: The effect of social modeling on treatment outcomes. *Current Directions in Psychological Science*, *25*(6), 438–43.
14. Mazzoni, G., Foan, L., Hyland, M.E., and Kirsch, I. 2010. The effects of observation and gender on psychogenic symptoms. *Health Psychology*, *29*, 181–85; Lorber, W., Mazzoni, G., and Kirsch, I. (2007). Illness by suggestion: Expectancy, modeling, and gender in the production of psychosomatic symptoms. *Annals of Behavioral Medicine*, *33*(1), 112–16.
15. Broderick, J.E., Kaplan-Liss, E., and Bass, E. (2011). Experimental induction of psychogenic illness in the context of a medical event and media exposure. *American Journal of Disaster Medicine*, *6*(3), 163.
16. Ditto, B., Byrne, N., Holly, C., and Balegh, S. (2014). Social contagion of vasovagal reactions in the blood collection clinic: A possible example of mass psychogenic illness. *Health Psychology*, *33*(7), 639.
17. Faasse, K., Yeom, B., Parkes, B., Kearney, J., and Petrie, K.J. (2018). The influence of social modeling, gender, and empathy on treatment side effects. *Annals of Behavioral Medicine*, *52*(7), 560–70.
18. Colloca, L., and Benedetti, F. (2009). Placebo analgesia induced by social observational learning. *Pain*, *144*(1–2), 28–34; Świder, K., and Bąbel, P. (2013). The effect of the sex of a model on nocebo hyperalgesia induced by social observational learning. *Pain*, *154*(8), 1312–17.
19. Benedetti, F., Durando, J., and Vighetti, S. (2014). Nocebo and placebo modulation of hypobaric hypoxia headache involves the cyclooxygenase-prostaglandins pathway. *Pain*, *155*(5), 921–28.
20. Caporael, L.R. (1976). Ergotism: The Satan loosed in Salem? *Science*, *192*(4234), 21–26.
21. Hatfield, E., Carpenter, M., and Rapson, R.L. (2014). Emotional contagion as a precursor to collective emotions. In C. von Scheve and M. Salmela (Eds.). *Collective Emotions: Perspectives from Psychology, Philosophy, and Sociology*, 108–22. Oxford: Oxford University Press. Some further details (including the true location of the mill) come from Baloh, R.W., and Bartholomew, R.E. (2020). A short history of spider, insect, and worm scares. In *Havana Syndrome: Mass Psychogenic Illness and the Real Story Behind the Embassy Mystery and Hysteria*, 151–66. Cham, Switzerland: Copernicus.
22. Baloh, R.W., and Bartholomew, R.E. (2020). A short history of spider, insect, and worm scares. In *Havana Syndrome: Mass Psychogenic Illness and the Real Story Behind the Embassy Mystery and Hysteria*, 151–66. Cham, Switzerland: Copernicus.
23. Talbot, M. (2002, June 2). Hysteria hysteria. *New York Times Magazine*. https://www.nytimes.com/2002/06/02/magazine/hysteria-hysteria.html.

24. Koran, L., and Oppmann, P. (2018, March 2). US embassy in Cuba to reduce staff indefinitely after "health attacks." CNN. https://edition.cnn.com/2018/03/02/politics/us-embassy-cuba-staff-reductions-attacks/index.html.
25. See the following for a full argument of Havana syndrome's psychogenic origins: Baloh, R.W., and Bartholomew, R.E. (2020). *Havana Syndrome: Mass Psychogenic Illness and the Real Story Behind the Embassy Mystery and Hysteria*. Cham, Switzerland: Copernicus. See also Stone, R. (2018). Sonic attack or mass paranoia. *Science*, doi:10.1126/science.aau5386; Hitt, J. (2019, January 6). The real story behind the Havana embassy mystery. *Vanity Fair*. https://www.vanityfair.com/news/2019/01/the-real-story-behind-the-havana-embassy-mystery; Leighton, T.G. (2018). Ultrasound in air—Guidelines, applications, public exposures, and claims of attacks in Cuba and China. *Journal of the Acoustical Society of America*, *144*(4), 2473–89; Bartholomew, R.E., and Baloh, R.W. (2020). Challenging the diagnosis of "Havana Syndrome" as a novel clinical entity. *Journal of the Royal Society of Medicine*, *113*(1), 7–11. The possibility that psychogenic contagion was amplifying and prolonging symptoms is discussed in National Academies of Sciences, Engineering, and Medicine (2020). *An Assessment of Illness in US Government Employees and Their Families at Overseas Embassies*. Although this report raises the possibility of a real weapon, other scientists remain unconvinced: see Vergano, D. (2020, December 7). Scientists are slamming a report saying microwave attacks could have caused "Havana syndrome" in US diplomats. BuzzFeed. https://www.buzzfeednews.com/article/danvergano/microwave-attacks-havana-syndrome-diplomats.
26. Entous, A., and Anderson, J.L. (2018, November 9). The mystery of the Havana syndrome. *New Yorker*. https://www.newyorker.com/magazine/2018/11/19/the-mystery-of-the-havana-syndrome.
27. Cited in Baloh, R.W., and Bartholomew, R.E. (2020). *Havana Syndrome: Mass Psychogenic Illness and the Real Story Behind the Embassy Mystery and Hysteria*, 21. Cham, Switzerland: Copernicus.
28. The telephone as a cause of ear troubles. (1889). *British Medical Journal*, *2*(1499), 671–72.
29. Rubin, G.J., Burns, M., and Wessely, S. (2014). Possible psychological mechanisms for "wind turbine syndrome": On the windmills of your mind. *Noise and Health*, *16*(69), 116.
30. Andrianome, S., De Seze, R., Braun, A., and Selmaoui, B. (2018). Descriptive self-reporting survey of people with idiopathic environmental intolerance attributed to electromagnetic fields (IEI-EMF): Similarities and comparisons with previous studies. *Journal of Public Health*, *26*(4), 461–73.
31. Rubin, G.J., Hahn, G., Everitt, B.S., Cleare, A.J., and Wessely, S. (2006). Are some people sensitive to mobile phone signals? Within participants double blind randomised provocation study. *British Medical Journal*, *332*(7546), 886–91.
32. Verrender, A., Loughran, S.P., Dalecki, A., Freudenstein, F., and Croft, R.J. (2018). Can explicit suggestions about the harmfulness of EMF exposure exacerbate a nocebo response in healthy controls? *Environmental Research*, *166*, 409–17.
33. Nyhan, B., and Reifler, J. (2015). Does correcting myths about the flu vaccine work? An experimental evaluation of the effects of corrective information. *Vaccine*, *33*(3), 459–64.
34. Nichol, K.L., Margolis, K.L., Lind, A., Murdoch, M., McFadden, R., Hauge, M., . . . and Drake, M. (1996). Side effects associated with influenza vaccination in healthy working adults: A randomized, placebo-controlled trial. *Archives of Internal Medicine*, *156*(14), 1546–50; World Health Organization (2012). Information sheet: Observed rate of vaccine reactions: influenza vaccine. https://www.who.int/vaccine_safety/initiative/tools/Influenza_Vaccine_rates_information_sheet.pdf?ua=1.
35. CDC. Misconceptions about seasonal flu and flu vaccines. https://www.cdc.gov/flu/prevent/misconceptions.htm.

36. World Health Organization. (2012). Information sheet: Observed rate of vaccine reactions: influenza vaccine. https://www.who.int/vaccine_safety/initiative/tools/Influenza_Vaccine_rates_information_sheet.pdf?ua=1; Tosh, P.K., Boyce, T.G., and Poland, G.A. (2008). Flu myths: Dispelling the myths associated with live attenuated influenza vaccine. *Mayo Clinic Proceedings 83*(1), 77–84.
37. Huang, W.T., Hsu, C.C., Lee, P.I., and Chuang, J.H. (2010). Mass psychogenic illness in nationwide in-school vaccination for pandemic influenza A (H1N1) 2009, Taiwan, November 2009–January 2010. *Eurosurveillance*, *15*(21), 19575.
38. Simas, C., Munoz, N., Arregoces, L., and Larson, H.J. (2019). HPV vaccine confidence and cases of mass psychogenic illness following immunization in Carmen de Bolivar, Colombia. *Human Vaccines and Immunotherapeutics*, *15*(1), 163–66.
39. Matthews, A., Herrett, E., Gasparrini, A., Van Staa, T., Goldacre, B., Smeeth, L., and Bhaskaran, K. (2016). Impact of statin related media coverage on use of statins: Interrupted time series analysis with UK primary care data. *BMJ*, *353*, *i*3283. doi: 10.1136/bmj.i3283.
40. See, for example, Rogers, L. (2015, November 3). Crippled by statins. *Daily Mail*. https://www.dailymail.co.uk/health/article-3300937/Crippled-statins-Cholesterol-busting-drugs-left-David-wheelchair-doctors-insisted-taking-them.html.
41. Finegold, J.A., Manisty, C.H., Goldacre, B., Barron, A.J., and Francis, D. P. (2014). What proportion of symptomatic side effects in patients taking statins are genuinely caused by the drug? Systematic review of randomized placebo-controlled trials to aid individual patient choice. *European Journal of Preventive Cardiology*, *21*(4), 464–74.
42. Newman, C.B., Preiss, D., Tobert, J.A., Jacobson, T.A., Page, R.L., Goldstein, L.B., . . . and Duell, P.B. (2019). Statin safety and associated adverse events: A scientific statement from the American Heart Association. *Arteriosclerosis, Thrombosis, and Vascular Biology*, *39*(2), e38–e81.
43. Khan, S., Holbrook, A., and Shah, B.R. (2018). Does Googling lead to statin intolerance? *International Journal of Cardiology*, *262*, 25–27.
44. Singh, P., Arora, A., Strand, T.A., Leffler, D.A., Catassi, C., Green, P.H., . . . and Makharia, G.K. (2018). Global prevalence of celiac disease: Systematic review and meta-analysis. *Clinical Gastroenterology and Hepatology*, *16*(6), 823–36.
45. https://www.nhs.uk/conditions/coeliac-disease.
46. Cianferoni, A. (2016). Wheat allergy: Diagnosis and management. *Journal of Asthma and Allergy*, *9*, 13.
47. Servick, K. (2018, May). The war on gluten. *Science*. https://www.sciencemag.org/news/2018/05/what-s-really-behind-gluten-sensitivity.
48. Molina-Infante, J., and Carroccio, A. (2017). Suspected nonceliac gluten sensitivity confirmed in few patients after gluten challenge in double-blind, placebo-controlled trials. *Clinical Gastroenterology and Hepatology*, *15*(3), 339–48. See the following for a separate meta-analysis showing a large nocebo effect: Lionetti, E., Pulvirenti, A., Vallorani, M., Catassi, G., Verma, A. K., Gatti, S., and Catassi, C. (2017). Re-challenge studies in non-celiac gluten sensitivity: A systematic review and meta-analysis. *Frontiers in Physiology*, *8*, 621. The role of expectation in gluten sensitivity is described in the following: Petrie, K.J., and Rief, W. (2019). Psychobiological mechanisms of placebo and nocebo effects: Pathways to improve treatments and reduce side effects. *Annual Review of Psychology*, *70*, 599–625.

    The following link contains the British Nutrition Foundation's interpretation of the study: https://www.nutrition.org.uk/bnfevents/events/252-nutritionscience/researchspotlight/1043-2017issue3.html.

49. Croall, I.D., Trott, N., Rej, A., Aziz, I., O'Brien, D.J., George, H.A., . . . and Hadjivassiliou, M. (2019). A population survey of dietary attitudes towards gluten. *Nutrients, 11*(6), 1276.
50. Unalp-Arida, A., Ruhl, C.E., Brantner, T.L., Everhart, J.E., and Murray, J.A. (2017). Less hidden celiac disease but increased gluten avoidance without a diagnosis in the United States: Findings from the National Health and Nutrition Examination Surveys from 2009 to 2014. *Mayo Clinic Proceedings 92*(1), 30–38; Cabrera-Chávez, F., Dezar, G.V., Islas-Zamorano, A.P., Espinoza-Alderete, J.G., Vergara-Jiménez, M.J., Magaña-Ordorica, D., and Ontiveros, N. (2017). Prevalence of self-reported gluten sensitivity and adherence to a gluten-free diet in Argentinian adult population. *Nutrients, 9*(1), 81.
51. Crichton, F., Dodd, G., Schmid, G., Gamble, G., and Petrie, K.J. (2014). Can expectations produce symptoms from infrasound associated with wind turbines? *Health Psychology, 33*(4), 360; Crichton, F., Chapman, S., Cundy, T., and Petrie, K.J. (2014). The link between health complaints and wind turbines: Support for the nocebo expectations hypothesis. *Frontiers in Public Health, 2*, 220.
52. Crichton, F., and Petrie, K.J. (2015). Health complaints and wind turbines: The efficacy of explaining the nocebo response to reduce symptom reporting. *Environmental Research, 140*, 449–55.
53. Framing can also help. See, for instance, Mao, A., Barnes, K., Sharpe, L., Geers, A.L., Helfer, S.G., Faasse, K., and Colagiuri, B. (2021). Using positive attribute framing to attenuate nocebo side effects: A cybersickness study. *Annals of Behavioral Medicine*. doi: 10.1093/abm/kaaa115.

## Chapter 5: Faster, Stronger, Fitter

1. Voet, W. (2001). *Breaking the Chain*, 104. London: Yellow Jersey.
2. Bannister, R. (2014). *Twin Tracks: The Autobiography*, Kindle edition, location 828. London: Robson Press.
3. https://olympics.com/tokyo-2020/en/news/amp/eliud-kipchoge-s-unstoppable-marathon-mindset-the-olympics-and-world-records.
4. Gonzalez, R. (2019, October 14). How Eliud Kipchoge pulled off his epic, sub-2-hour marathon. *Wired*. https://www.wired.com/story/how-eliud-kipchoge-pulled-off-his-epic-sub-2-hour-marathon.
5. Giulio, C.D., Daniele, F., and Tipton, C.M. (2006). Angelo Mosso and muscular fatigue: 116 years after the first Congress of Physiologists: IUPS commemoration. *Advances in Physiology Education, 30*(2), 51–57.
6. Noakes, T.D. (2012). Fatigue is a brain-derived emotion that regulates the exercise behavior to ensure the protection of whole body homeostasis. *Frontiers in Physiology, 3*, 82.
7. Cairns, S. P. (2006). Lactic acid and exercise performance. *Sports Medicine, 36*(4), 279–91. See also https://www.livescience.com/lactic-acid.html.
8. Corbett, J., Barwood, M.J., Ouzounoglou, A., Thelwell, R., and Dicks, M. (2012). Influence of competition on performance and pacing during cycling exercise. *Medicine and Science in Sports and Exercise, 44*(3), 509–15.
9. Marcora, S.M., Staiano, W., and Manning, V. (2009). Mental fatigue impairs physical performance in humans. *Journal of Applied Physiology, 106*(3), 857–64.
10. For a thorough discussion of the traditional model of fatigue, and the need to separate the psychological sense of effort from the physiological changes, see Noakes, T.D. (2012). The Central Governor Model in 2012: Eight new papers deepen our understanding of

the regulation of human exercise performance. *British Journal of Sports Medicine*, *46*, 1–3. There has been controversy over the exact formulation of the psychobiological theory of fatigue, though the description in the text describes the common features. See Venhorst, A., Micklewright, D., and Noakes, T.D. (2018). Towards a three-dimensional framework of centrally regulated and goal-directed exercise behaviour: A narrative review. *British Journal of Sports Medicine*, *52*(15), 957–66.

11. For some direct evidence of this part of the process, see Piedimonte, A., Benedetti, F., and Carlino, E. (2015). Placebo-induced decrease in fatigue: Evidence for a central action on the preparatory phase of movement. *European Journal of Neuroscience*, *41*(4), 492–97.
12. Morton, R.H. (2009). Deception by manipulating the clock calibration influences cycle ergometer endurance time in males. *Journal of Science and Medicine in Sport*, *12*, 332–37.
13. Stone, M., Thomas, K., Wilkinson, M., Jones, A., St. Clair Gibson, A., and Thompson, K. (2012). Effects of deception on exercise performance: Implications for determinants of fatigue in humans. *Medicine and Science in Sports and Exercise*, *44*(3), 534–41.
14. Castle, P.C., Maxwell, N., Allchorn, A., Mauger, A.R., and White, D.K. (2012). Deception of ambient and body core temperature improves self paced cycling in hot, humid conditions. *European Journal of Applied Physiology*, *112*(1), 377–85.
15. Iodice, P., Porciello, G., Bufalari, I., Barca, L., and Pezzulo, G. (2019). An interoceptive illusion of effort induced by false heart-rate feedback. *Proceedings of the National Academy of Sciences*, *116*(28), 13897–902.
16. McMorris, T., Barwood, M., and Corbett, J. (2018). Central fatigue theory and endurance exercise: Toward an interoceptive model. *Neuroscience and Biobehavioral Reviews*, *93*, 93–107; Holgado, D., and Sanabria, D. (2020). Does self-paced exercise depend on executive processing? A narrative review of the current evidence. *International Review of Sport and Exercise Psychology*, 1–24; Hyland-Monks, R., Cronin, L., McNaughton, L., and Marchant, D. (2018). The role of executive function in the self-regulation of endurance performance: A critical review. In *Progress in Brain Research*, *240*, 353–70.
17. Broelz, E.K., Wolf, S., Schneeweiss, P., Niess, A.M., Enck, P., and Weimer, K. (2018). Increasing effort without noticing: A randomized controlled pilot study about the ergogenic placebo effect in endurance athletes and the role of supplement salience. *PLoS One*, *13*(6), e0198388.
18. Pollo, A., Carlino, E., and Benedetti, F. (2008). The top-down influence of ergogenic placebos on muscle work and fatigue. *European Journal of Neuroscience*, *28*(2), 379–88.
19. Hurst, P., Schipof-Godart, L., Szabo, A., Raglin, J., Hettinga, F., Roelands, B., . . . and Beedie, C. (2020). The placebo and nocebo effect on sports performance: A systematic review. *European Journal of Sport Science*, *20*(3), 279–92.
20. Ibid.
21. Montes, J., Wulf, G., and Navalta, J.W. (2018). Maximal aerobic capacity can be increased by enhancing performers' expectancies. *Journal of Sports Medicine and Physical Fitness*, *58*(5), 744–49.
22. Stoate, I., Wulf, G., and Lewthwaite, R. (2012). Enhanced expectancies improve movement efficiency in runners. *Journal of Sports Sciences*, *30*(8), 815–23.
23. Turnwald, B.P., Goyer, J.P., Boles, D.Z., Silder, A., Delp, S.L., and Crum, A.J. (2019). Learning one's genetic risk changes physiology independent of actual genetic risk. *Nature Human Behaviour*, *3*(1), 48–56.
24. Saito, T., Barreto, G., Saunders, B., and Gualano, B. (2020). Is open-label placebo a new ergogenic aid? A commentary on existing studies and guidelines for future research. *Sports Medicine*, *50*(7), 1231–32. See also Broelz, E.K., Wolf, S., Schneeweiss, P., Niess, A.M., Enck, P., and Weimer, K. (2018). Increasing effort without noticing: A randomized con-

trolled pilot study about the ergogenic placebo effect in endurance athletes and the role of supplement salience. *PLoS One*, *13*(6), e0198388.

25. Giles, G.E., Cantelon, J.A., Eddy, M.D., Brunyé, T.T., Urry, H.L., Taylor, H.A., . . . and Kanarek, R.B. (2018). Cognitive reappraisal reduces perceived exertion during endurance exercise. *Motivation and Emotion*, *42*(4), 482–96. Some of the advice given here is based on an interview with Giles, and my own experience of practicing cognitive reappraisal. For another example of cognitive reappraisal, see Arthur, T.G., Wilson, M.R., Moore, L.J., Wylie, L.J., and Vine, S.J. (2019). Examining the effect of challenge and threat states on endurance exercise capabilities. *Psychology of Sport and Exercise*, *44*, 51–59. And see the following for a discussion of emotional intelligence and its relation to the psychological basis of fatigue: Rubaltelli, E., Agnoli, S., and Leo, I. (2018). Emotional intelligence impact on half marathon finish times. *Personality and Individual Differences*, *128*, 107–12.
26. Orvidas, K., Burnette, J.L., and Russell, V.M. (2018). Mindsets applied to fitness: Growth beliefs predict exercise efficacy, value and frequency. *Psychology of Sport and Exercise*, *36*, 156–61.
27. Morris, J.N., Heady, J.A., Raffle, P.A.B., Roberts, C.G., and Parks, J.W. (1953). Coronary heart-disease and physical activity of work. *Lancet*, *262*(6796), 1111–20; Kuper, S. (2009, September 12). The man who invented exercise. *Financial Times*. https://www.ft.com/content/e6ff90ea-9da2-11de-9f4a-00144feabdc0; Paffenbarger Jr., R.S., Blair, S.N., and Lee, I.M. (2001). A history of physical activity, cardiovascular health and longevity: The scientific contributions of Jeremy N. Morris, DSc, DPH, FRCP. *International Journal of Epidemiology*, *30*(5), 1184–92.
28. Source: https://sites.google.com/site/compendiumofphysicalactivities/home. See also Wilson, C. (2010). The truth about exercise. *New Scientist*, *205*(2742), 34–37.
29. Patterson, R., Webb, E., Millett, C., and Laverty, A.A. (2018). Physical activity accrued as part of public transport use in England. *Journal of Public Health*, *41(2)*, 222–30.
30. Crum, A.J., and Langer, E.J. (2007). Mind-set matters: Exercise and the placebo effect. *Psychological Science*, *18*(2), 165–71.
31. Zahrt, O.H., and Crum, A.J. (2017). Perceived physical activity and mortality: Evidence from three nationally representative US samples. *Health Psychology*, *36*(11), 1017. A similar study, looking at people's health complaints: Baceviciene, M., Jankauskiene, R., and Emeljanovas, A. (2019). Self-perception of physical activity and fitness is related to lower psychosomatic health symptoms in adolescents with unhealthy lifestyles. *BMC Public Health*, *19*(1), 980.
32. Lindheimer, J.B., O'Connor, P.J., and Dishman, R.K. (2015). Quantifying the placebo effect in psychological outcomes of exercise training: A meta-analysis of randomized trials. *Sports Medicine*, *45*(5), 693–711; Jones, M.D., Valenzuela, T., Booth, J., Taylor, J.L., and Barry, B.K. (2017). Explicit education about exercise-induced hypoalgesia influences pain responses to acute exercise in healthy adults: A randomized controlled trial. *Journal of Pain*, *18*(11), 1409–16; Vaegter, H.B., Thinggaard, P., Madsen, C.H., Hasenbring, M., and Thorlund, J.B. (2020). Power of words: Influence of preexercise information on hypoalgesia after exercise-randomized controlled trial. *Medicine and Science in Sports and Exercise*, *52*(11), 2373–79.
33. Zahrt, O.H., and Crum, A.J. (2019). Effects of physical activity recommendations on mindset, behavior and perceived health. *Preventive Medicine Reports*, 101027.
34. Wen, C.P., Wai, J.P.M., Tsai, M.K., Yang, Y.C., Cheng, T.Y.D., Lee, M.C., . . . and Wu, X. (2011). Minimum amount of physical activity for reduced mortality and extended life expectancy: A prospective cohort study. *Lancet*, *378*(9798), 1244–53. See also Curfman, G. (2015, December 8). Exercise: You may need less than you think. https://www.health.harvard.edu/blog/how-much-exercise-do-you-really-need-less-than-you-think-201512088770.

35. Prichard, I., Kavanagh, E., Mulgrew, K.E., Lim, M.S., and Tiggemann, M. (2020). The effect of Instagram #fitspiration images on young women's mood, body image, and exercise behaviour. *Body Image*, *33*, 1–6. See also Robinson, L., Prichard, I., Nikolaidis, A., Drummond, C., Drummond, M., and Tiggemann, M. (2017). Idealised media images: The effect of fitspiration imagery on body satisfaction and exercise behaviour. *Body Image*, *22*, 65–71.
36. Phelps, M., with Abrahamson, A. (2008). *No Limits: The Will to Succeed*, 8. New York: Free Press. Cited in Moran, A., Campbell, M., Holmes, P., and MacIntyre, T. (2012). Mental imagery, action observation and skill learning. In N.J. Hodges and A.M. Williams (Eds.). *Skill Acquisition in Sport: Research, Theory and Practice*, 94. London: Routledge.
37. Moran, A., Campbell, M., Holmes, P., and MacIntyre, T. (2012). Mental imagery, action observation and skill learning. In N.J. Hodges and A.M. Williams (Eds.). *Skill Acquisition in Sport: Research, Theory and Practice*, 94. London: Routledge.
    See also Slimani, M., Tod, D., Chaabene, H., Miarka, B., and Chamari, K. (2016). Effects of mental imagery on muscular strength in healthy and patient participants: A systematic review. *Journal of Sports Science and Medicine*, *15*(3), 434.
38. Yao, W.X., Ranganathan, V.K., Allexandre, D., Siemionow, V., and Yue, G.H. (2013). Kinesthetic imagery training of forceful muscle contractions increases brain signal and muscle strength. *Frontiers in Human Neuroscience*, *7*, 561. See the following for a comparison of physical and mental practice, and various combinations of both styles of training: Reiser, M., Büsch, D., and Munzert, J. (2011). Strength gains by motor imagery with different ratios of physical to mental practice. *Frontiers in Psychology*, *2*, 194.
39. While this has been the view for many decades, the latest evidence suggests that the size of our muscles and muscular strength are largely independent. Loenneke, J.P., Buckner, S.L., Dankel, S.J., and Abe, T. (2019). Exercise-induced changes in muscle size do not contribute to exercise-induced changes in muscle strength. *Sports Medicine*, *49*(7), 987–91.
40. Ridderinkhof, K.R., and Brass, M. (2015). How kinesthetic motor imagery works: A predictive-processing theory of visualization in sports and motor expertise. *Journal of Physiology—Paris*, *109*(1–3), 53–63. See the following for a discussion of its relation to the psychobiological model of exercise: Slimani, M., Tod, D., Chaabene, H., Miarka, B., and Chamari, K. (2016). Effects of mental imagery on muscular strength in healthy and patient participants: A systematic review. *Journal of Sports Science and Medicine*, *15*(3), 434.
41. Lebon, F., Collet, C., and Guillot, A. (2010). Benefits of motor imagery training on muscle strength. *Journal of Strength and Conditioning Research*, *24*(6), 1680–87.
42. Clark, B.C., Mahato, N.K., Nakazawa, M., Law, T.D., and Thomas, J.S. (2014). The power of the mind: The cortex as a critical determinant of muscle strength/weakness. *Journal of Neurophysiology*, *112*(12), 3219–26.
43. See, for example, Najafabadi, M.G., Memari, A.H., Kordi, R., Shayestehfar, M., and Eshghi, M.A. (2017). Mental training can improve physical activity behavior in adolescent girls. *Journal of Sport and Health Science*, *6*(3), 327–32; Cooke, L.M., Duncan, L.R., Deck, S.J., Hall, C.R., and Rodgers, W.M. (2020). An examination of changes in exercise identity during a mental imagery intervention for female exercise initiates. *International Journal of Sport and Exercise Psychology*, *18*(4), 534–50; Robin, N., Toussaint, L., Coudevylle, G.R., Ruart, S., Hue, O., and Sinnapah, S. (2018). Text messages promoting mental imagery increase self-reported physical activity in older adults: A randomized controlled study. *Journal of Aging and Physical Activity*, *26*(3), 462–70.
44. Newcomb, A. (2012, August 1). Super strength: Daughter rescues dad trapped under car. ABC News. https://abcnews.go.com/US/superhero-woman-lifts-car-off-dad/story?id=16907591#.UMay9Hfeba4. See also Hadhazy, A. (2016, May 2). How it's possible

for an ordinary person to lift a car. BBC Future. https://www.bbc.com/future/article/20160501-how-its-possible-for-an-ordinary-person-to-lift-a-car.

45. Oregon man pinned under 3,000-pound tractor saved by teen daughters. (2013, April 11). Fox News. https://www.foxnews.com/us/oregon-man-pinned-under-3000-pound-tractor-saved-by-teen-daughters; Septuagenarian superhero? Man lifts car off son-in-law. (2013, July 22). NPR. https://www.npr.org/2013/07/22/204444515/septuagenarian-superhero-man-lifts-car-off-son-in-law.
46. Liptak, A. (2015, August 30). The Incredible Hulk was inspired by a woman saving her baby. *Gizmodo*. https://io9.gizmodo.com/the-incredible-hulk-was-inspired-by-a-woman-saving-her-1727562968.
47. Evans, D.R., Boggero, I.A., and Segerstrom, S.C. (2016). The nature of self-regulatory fatigue and "ego depletion": Lessons from physical fatigue. *Personality and Social Psychology Review*, *20*(4), 291–310.

## Chapter 6: The Food Paradox

1. Calorie content: avocado toast (501 cal); smoothie (209 cal); tuna niçoise salad (455 cal); orange juice (105 cal); chicken and asparagus braise (480 cal); fruit-and-nut granola bar (279 cal). Sources: www.bbcgood.food.com, www.pret.co.uk.
2. Calorie content: croissant (291 cal); hot chocolate (260 cal); spaghetti alla puttanesca (495 cal), and fruit salad (111 cal); fish pie (455 cal) and salad (20 cal); two mini donuts (110 cal). Sources: www.pret.co.uk, www.bbcgoodfood.com, www.sainsburys.co.uk.
3. In the discussion of Henry Molaison's life that follows, I am indebted to Corkin, S. (2014). *Permanent Present Tense*. London: Penguin.
4. Ibid., 210.
5. For descriptions of this experiment and its implications for the role of memory in appetite, see Rozin, P., Dow, S., Moscovitch, M., and Rajaram, S. (1998). What causes humans to begin and end a meal? A role for memory for what has been eaten, as evidenced by a study of multiple meal eating in amnesic patients. *Psychological Science*, *9*(5), 392–96; and Higgs, S. (2005). Memory and its role in appetite regulation. *Physiology and Behavior*, *85*(1), 67–72.
6. Berthoud, H.R. (2008). Vagal and hormonal gut-brain communication: From satiation to satisfaction. *Neurogastroenterology and Motility*, *20*, 64–72.
7. Desai, A.J., Dong, M., Harikumar, K.G., and Miller, L.J. (2016). Cholecystokinin-induced satiety, a key gut servomechanism that is affected by the membrane microenvironment of this receptor. *International Journal of Obesity Supplements*, *6*(1), S22–S27.
8. Martin, A.A., Davidson, T.L., and McCrory, M.A. (2018). Deficits in episodic memory are related to uncontrolled eating in a sample of healthy adults. *Appetite*, *124*, 33–42.
9. Higgs, S. (2002). Memory for recent eating and its influence on subsequent food intake. *Appetite*, *39*(2), 159–66. Higgs has also found that the effect of memory depends on someone's overall level of inhibition. See Higgs, S., Williamson, A.C., and Attwood, A.S. (2008). Recall of recent lunch and its effect on subsequent snack intake. *Physiology and Behavior*, *94*(3), 454–62.
10. Brunstrom, J.M., Burn, J.F., Sell, N.R., Collingwood, J.M., Rogers, P.J., Wilkinson, L.L., . . . and Ferriday, D. (2012). Episodic memory and appetite regulation in humans. *PLoS One*, *7*(12), e50707.
11. Brown, S.D., Duncan, J., Crabtree, D., Powell, D., Hudson, M., and Allan, J.L. (2020). We are what we (think we) eat: The effect of expected satiety on subsequent calorie consumption. *Appetite*, 152, 104717.

12. Higgs, S., and Woodward, M. (2009). Television watching during lunch increases afternoon snack intake of young women. *Appetite*, *52*(1), 39–43; Higgs, S. (2015). Manipulations of attention during eating and their effects on later snack intake. *Appetite*, *92*, 287–94. See the following for a review of these findings: Higgs, S., and Spetter, M.S. (2018). Cognitive control of eating: The role of memory in appetite and weight gain. *Current Obesity Reports*, *7*(1), 50–59.
13. Brunstrom, J.M., Brown, S., Hinton, E.C., Rogers, P.J., and Fay, S.H. (2011). "Expected satiety" changes hunger and fullness in the inter-meal interval. *Appetite*, *56*(2), 310–15.
14. Cornil, Y. (2017). Mind over stomach: A review of the cognitive drivers of food satiation. *Journal of the Association for Consumer Research*, 2(4), 419–29.
15. Finkelstein, S.R., and Fishbach, A. (2010). When healthy food makes you hungry. *Journal of Consumer Research*, *37*(3), 357–67.
16. Abizaid, A., and Horvath, T.L. (2012). Ghrelin and the central regulation of feeding and energy balance. *Indian Journal of Endocrinology and Metabolism*, *16* (Suppl 3), S617.
17. Crum, A.J., Corbin, W.R., Brownell, K.D., and Salovey, P. (2011). Mind over milkshakes: Mindsets, not just nutrients, determine ghrelin response. *Health Psychology*, *30*(4), 424. See the following for a peer commentary on the results and their potential implications for weight management: Tomiyama, A.J., and Mann, T. (2011). Commentary on Crum, Corbin, Brownell, and Salovey (2011). *Health Psychology*, *30*(4), 430–31.
18. I spoke to Alia Crum for the following article: Robson, D. (2018). Mind over matter. *New Scientist*, *239*(3192), 28–32.
19. Veldhuizen, M.G., Nachtigal, D.J., Flammer, L.J., de Araujo, I.E., and Small, D.M. (2013). Verbal descriptors influence hypothalamic response to low-calorie drinks. *Molecular Metabolism*, *2*(3), 270–80.
20. Cassady, B.A., Considine, R.V., and Mattes, R.D. (2012). Beverage consumption, appetite, and energy intake: What did you expect? *American Journal of Clinical Nutrition*, *95*(3), 587–93.
21. Yeomans, M.R., Re, R., Wickham, M., Lundholm, H., and Chambers, L. (2016). Beyond expectations: The physiological basis of sensory enhancement of satiety. *International Journal of Obesity*, *40*(11), 1693–98; Zhu, Y., Hsu, W.H., and Hollis, J.H. (2013). The impact of food viscosity on eating rate, subjective appetite, glycemic response and gastric emptying rate. *PLoS One*, *8*(6), e67482.
22. Hallberg, L., Björn-Rasmussen, E., Rossander, L., and Suwanik, R. (1977). Iron absorption from Southeast Asian diets. II. Role of various factors that might explain low absorption. *American Journal of Clinical Nutrition*, *30*(4), 539–48.
23. Björn-Rasmussen, E., Halberg, L., Magnusson, B., Rossander, L., Svanberg, B., and Arvidsson, B. (1976). Measurement of iron absorption from composite meals. *American Journal of Clinical Nutrition*, *29*(7), 772–78; Hallberg, L., Björn-Rasmussen, E., Rossander, L., and Suwanik, R. (1977). Iron absorption from Southeast Asian diets. II. Role of various factors that might explain low absorption. *American Journal of Clinical Nutrition*, *30*(4), 539–48. For a more recent analysis of these results, see Satter, E. (2007). Eating competence: Definition and evidence for the Satter Eating Competence model. *Journal of Nutrition Education and Behavior*, *39*(5), S142–S153.
24. Todes, Daniel P. (2014, November 21). Ivan Pavlov in 22 surprising facts. https://blog.oup.com/2014/11/ivan-pavlov-surprising-facts.
25. Jonas, W.B., Crawford, C., Colloca, L., Kaptchuk, T.J., Moseley, B., Miller, F.G., . . . and Meissner, K. (2015). To what extent are surgery and invasive procedures effective beyond a placebo response? A systematic review with meta-analysis of randomised, sham controlled trials. *BMJ Open*, *5*(12), e009655.

26. https://www.who.int/news-room/fact-sheets/detail/obesity-and-overweight.
27. Carels, R.A., Harper, J., and Konrad, K. (2006). Qualitative perceptions and caloric estimations of healthy and unhealthy foods by behavioral weight loss participants. *Appetite*, *46*(2), 199–206.
28. Suher, J., Raghunathan, R., and Hoyer, W.D. (2016). Eating healthy or feeling empty? How the "healthy = less filling" intuition influences satiety. *Journal of the Association for Consumer Research*, *1*(1), 26–40.
29. Briers, B., Huh, Y.E., Chan, E., and Mukhopadhyay, A. (2020). The unhealthy = tasty belief is associated with BMI through reduced consumption of vegetables: A cross-national and mediational analysis. *Appetite*, *150*, 104639. See also Cooremans, K., Geuens, M., and Pandelaere, M. (2017). Cross-national investigation of the drivers of obesity: Reassessment of past findings and avenues for the future. *Appetite*, *114*, 360–67.
30. Raghunathan, R., Naylor, R.W., and Hoyer, W.D. (2006). The unhealthy = tasty intuition and its effects on taste inferences, enjoyment, and choice of food products. *Journal of Marketing*, *70*(4), 170–84.
31. Turnwald, B.P., Jurafsky, D., Conner, A., and Crum, A.J. (2017). Reading between the menu lines: Are restaurants' descriptions of "healthy" foods unappealing? *Health Psychology*, *36*(11), 1034.
32. Turnwald, B.P., Boles, D.Z., and Crum, A.J. (2017). Association between indulgent descriptions and vegetable consumption: Twisted carrots and dynamite beets. *JAMA Internal Medicine*, *177*(8), 1216–18; Turnwald, B.P., Bertoldo, J.D., Perry, M.A., Policastro, P., Timmons, M., Bosso, C., . . . and Gardner, C.D. (2019). Increasing vegetable intake by emphasizing tasty and enjoyable attributes: A randomized controlled multisite intervention for taste-focused labeling. *Psychological Science*, *30*(11), 1603–15.
33. Fay, S.H., Hinton, E.C., Rogers, P.J., and Brunstrom, J.M. (2011). Product labelling can confer sustained increases in expected and actual satiety. *Appetite*, *57*(2), 557.
34. Cheon, B.K., and Hong, Y.Y. (2017). Mere experience of low subjective socioeconomic status stimulates appetite and food intake. *Proceedings of the National Academy of Sciences*, *114*(1), 72–77.
35. Sim, A.Y., Lim, E.X., Leow, M.K., and Cheon, B.K. (2018). Low subjective socioeconomic status stimulates orexigenic hormone ghrelin: A randomised trial. *Psychoneuroendocrinology*, *89*, 103–12.
36. Brunstrom, J.M., Brown, S., Hinton, E.C., Rogers, P.J., and Fay, S.H. (2011). "Expected satiety" changes hunger and fullness in the inter-meal interval. *Appetite*, *56*(2), 310–15.
37. https://www.health.harvard.edu/staying-healthy/the-hidden-dangers-of-protein-powders.
38. Mandel, N., and Brannon, D. (2017). Sugar, perceived healthfulness, and satiety: When does a sugary preload lead people to eat more? *Appetite*, *114*, 338–49.
39. Yeomans, M.R. (2015). Cued satiety: How consumer expectations modify responses to ingested nutrients. *Nutrition Bulletin*, *40*(2), 100–3.
40. Kuijer, R.G., and Boyce, J.A. (2014). Chocolate cake. Guilt or celebration? Associations with healthy eating attitudes, perceived behavioural control, intentions and weight loss. *Appetite*, *74*, 48–54.
41. Cornil, Y., and Chandon, P. (2016). Pleasure as a substitute for size: How multisensory imagery can make people happier with smaller food portions. *Journal of Marketing Research*, *53*(5), 847–64. The following paper found a similar effect with food writing—the richer the description of a cake, the less people wanted to eat, and the more satisfied they felt after eating it: Policastro, P., Harris, C., and Chapman, G. (2019). Tasting with your eyes: Sensory description substitutes for portion size. *Appetite*, *139*, 42–49.

42. Morewedge, C.K., Huh, Y.E., and Vosgerau, J. (2010). Thought for food: Imagined consumption reduces actual consumption. *Science*, *330*(6010), 1530–33.
43. There is even evidence that the anticipation of food can alter the ghrelin suppression, after eating: Ott, V., Friedrich, M., Zemlin, J., Lehnert, H., Schultes, B., Born, J., and Hallschmid, M. (2012). Meal anticipation potentiates postprandial ghrelin suppression in humans. *Psychoneuroendocrinology*, *37*(7), 1096–100.
44. Bosworth, M.L., Ferriday, D., Lai, S.H.S., Godinot, N., Martin, N., Martin, A.A., . . . and Brunstrom, J.M. (2016). Eating slowly increases satiety and promotes memory of a larger portion size during the inter-meal interval. *Appetite*, *100*(101), 225.
45. Raghunathan, R., Naylor, R.W., and Hoyer, W.D. (2006). The unhealthy = tasty intuition and its effects on taste inferences, enjoyment, and choice of food products. *Journal of Marketing*, *70*(4), 170–84.
46. Briers, B., Huh, Y.E., Chan, E., and Mukhopadhyay, A. (2020). The unhealthy = tasty belief is associated with BMI through reduced consumption of vegetables: A cross-national and mediational analysis. *Appetite*, *150*, 104639.
47. Werle, C.O., Trendel, O., and Ardito, G. (2013). Unhealthy food is not tastier for everybody: The "healthy = tasty" French intuition. *Food Quality and Preference*, *28*(1), 116–21.
48. Rozin, P., Kabnick, K., Pete, E., Fischler, C., and Shields, C. (2003). The ecology of eating: Smaller portion sizes in France than in the United States help explain the French paradox. *Psychological Science*, *14*(5), 450–54.
49. World Health Organization. (2014). *Global Status Report on Noncommunicable Diseases 2014*.
50. Rozin, P., Fischler, C., Imada, S., Sarubin, A., and Wrzesniewski, A. (1999). Attitudes to food and the role of food in life in the USA, Japan, Flemish Belgium and France: Possible implications for the diet-health debate. *Appetite*, *33*(2), 163–80.

## Chapter 7: De-stressing Stress

1. Increase of heart-disease. (1872). *British Medical Journal*, *1*(586), 317.
2. Theodore Seward starts "Don't Worry" clubs. (1898, January 17). *The Gazette* (York, PA), 3; Don't Worry circles (1897, December 19). *New York Times*, 7.
3. Seward, T. (1898). *The Don't Worry Movement: A Wave of Spiritual Emancipation* (self-published).
4. James, W. (1902). *The Varieties of Religious Experience*, 94. New York: Longman.
5. James, W. (1963). *Pragmatism, and Other Essays*, 237. New York: Washington Square Press.
6. Wallis, C., Mehrtens, R., and Thompson, D. (1983). Stress: Can we cope? *Time*, *121*(23), 48–54.
7. https://www.merriam-webster.com/dictionary/stressed-out.
8. https://www.health.harvard.edu/staying-healthy/understanding-the-stress-response. See also Burrows, V.L. (2015). The medicalization of stress: Hans Selye and the transformation of the postwar medical marketplace. Unpublished PhD thesis, City University of New York. https://academicworks.cuny.edu/gc_etds/877.
9. The preceding paragraphs are indebted to Jackson, M. (2014). *Stress, Shock, and Adaptation in the Twentieth Century*, esp. ch. 1. Rochester, NY: University of Rochester Press; Burrows, V.L. (2015). The medicalization of stress: Hans Selye and the transformation of the postwar medical marketplace. Unpublished PhD thesis, City University of New York. See the following for a modern description of the physiological and mental changes caused by threat: Mendes, W.B., and Park, J. (2014). Neurobiological concomitants of motivational states. *Advances in Motivation Science*, *1*, 233–70.

10. Jamieson, J.P., Peters, B.J., Greenwood, E.J., and Altose, A.J. (2016). Reappraising stress arousal improves performance and reduces evaluation anxiety in classroom exam situations. *Social Psychological and Personality Science*, *7*(6), 579–87.
11. Jamieson, J.P., Mendes, W.B., Blackstock, E., and Schmader, T. (2010). Turning the knots in your stomach into bows: Reappraising arousal improves performance on the GRE. *Journal of Experimental Social Psychology*, *46*(1), 208–12.
12. Jamieson, J.P., Nock, M.K., and Mendes, W.B. (2012). Mind over matter: Reappraising arousal improves cardiovascular and cognitive responses to stress. *Journal of Experimental Psychology: General*, *141*(3), 417. Further interpretation (and information on recovery): Jamieson, J.P., Mendes, W.B., and Nock, M.K. (2013). Improving acute stress responses: The power of reappraisal. *Current Directions in Psychological Science*, *22*(1), 51–56. See also Mendes, W.B., and Park, J. (2014). Neurobiological concomitants of motivational states. *Advances in Motivation Science*, *1*, 233–70; Trotman, G.P., Williams, S.E., Quinton, M.L., and van Zanten, J.J.V. (2018). Challenge and threat states: Examining cardiovascular, cognitive and affective responses to two distinct laboratory stress tasks. *International Journal of Psychophysiology*, *126*, 42–51.
13. See the following for a thorough analysis of stress appraisals, the cardiovascular responses, and the link to performance: Behnke, M., and Kaczmarek, L.D. (2018). Successful performance and cardiovascular markers of challenge and threat: A meta-analysis. *International Journal of Psychophysiology*, *130*, 73–79.
14. Crum, A.J., Salovey, P., and Achor, S. (2013). Rethinking stress: The role of mindsets in determining the stress response. *Journal of Personality and Social Psychology*, *104*(4), 716.
15. Crum, A.J., Akinola, M., Martin, A., and Fath, S. (2017). The role of stress mindset in shaping cognitive, emotional, and physiological responses to challenging and threatening stress. *Anxiety, Stress, and Coping*, *30*(4), 379–95; John-Henderson, N.A., Rheinschmidt, M.L., and Mendoza-Denton, R. (2015). Cytokine responses and math performance: The role of stereotype threat and anxiety reappraisals. *Journal of Experimental Social Psychology*, *56*, 203–6.
16. See the following for a general description of the differences between "threat" and "challenge" states: Blascovich, J., and Mendes, W.B. (2010). Social psychophysiology and embodiment. In S.T. Fiske, D.T. Gilbert, and G. Lindzey (Eds.). *The Handbook of Social Psychology* (5th ed., 194–227). New York: Wiley.
17. Crum, A.J., Akinola, M., Martin, A., and Fath, S. (2017). The role of stress mindset in shaping cognitive, emotional, and physiological responses to challenging and threatening stress. *Anxiety, Stress, and Coping*, *30*(4), 379–95.
18. Akinola, M., Fridman, I., Mor, S., Morris, M.W., and Crum, A.J. (2016). Adaptive appraisals of anxiety moderate the association between cortisol reactivity and performance in salary negotiations. *PLoS One*, *11*(12), e0167977.
19. Smith, E.N., Young, M.D., and Crum, A.J. (2020). Stress, mindsets, and success in Navy SEALs special warfare training. *Frontiers in Psychology*, *10*, 2962.
20. Beltzer, M.L., Nock, M.K., Peters, B.J., and Jamieson, J.P. (2014). Rethinking butterflies: The affective, physiological, and performance effects of reappraising arousal during social evaluation. *Emotion*, *14*(4), 761.
21. Strack, J., Lopes, P.N., and Esteves, F. (2015). Will you thrive under pressure or burn out? Linking anxiety motivation and emotional exhaustion. *Cognition and Emotion*, *29*(4), 578–91. For further examples, see Kim, J., Shin, Y., Tsukayama, E., and Park, D. (2020). Stress mindset predicts job turnover among preschool teachers. *Journal of School Psychology*, *78*, 13–22; Keech, J.J., Cole, K.L., Hagger, M.S., and Hamilton, K. (2020). The association between stress mindset and physical and psychological wellbeing: Testing a stress beliefs model in police officers. *Psychology and Health*, *35*(11), 1306–25; Casper, A.,

Sonnentag, S., and Tremmel, S. (2017). Mindset matters: The role of employees' stress mindset for day-specific reactions to workload anticipation. *European Journal of Work and Organizational Psychology*, *26*(6), 798–810.

22. Keller, A., Litzelman, K., Wisk, L.E., Maddox, T., Cheng, E.R., Creswell, P.D., and Witt, W.P. (2012). Does the perception that stress affects health matter? The association with health and mortality. *Health Psychology*, *31*(5), 677. See the following for a near-exact replication of this result: Nabi, H., Kivimäki, M., Batty, G.D., Shipley, M.J., Britton, A., Brunner, E.J., . . . and Singh-Manoux, A. (2013). Increased risk of coronary heart disease among individuals reporting adverse impact of stress on their health: The Whitehall II prospective cohort study. *European Heart Journal*, *34*(34), 2697–705.
23. Szabo, A., and Kocsis, Á. (2017). Psychological effects of deep-breathing: The impact of expectancy-priming. *Psychology, Health and Medicine*, *22*(5), 564–69; Cregg, D.R., and Cheavens, J.S. (2020). Gratitude interventions: Effective self-help? A meta-analysis of the impact on symptoms of depression and anxiety. *Journal of Happiness Studies*, 22, 413–45.
24. Brady, S.T., Hard, B.M., and Gross, J.J. (2018). Reappraising test anxiety increases academic performance of first-year college students. *Journal of Educational Psychology*, *110*(3), 395.
25. The advice in this section is based on the following paper: Keech, J.J., Hagger, M.S., and Hamilton, K. (2019). Changing stress mindsets with a novel imagery intervention: A randomized controlled trial. *Emotion*, *21*(1), 123–36. See also the following resources: http://socialstresslab.wixsite.com/urochester/research; https://mbl.stanford.edu/interventions/rethink-stress.
26. Jentsch, V.L., and Wolf, O.T. (2020). The impact of emotion regulation on cardiovascular, neuroendocrine and psychological stress responses. *Biological Psychology*, 107893.
27. King, B.J. (2008). *Pressure Is Privilege*, 102–3. New York: LifeTime.
28. I interviewed Mauss for the following article: Robson, D. (2018, December 18). Why the quickest route to happiness may be to do nothing. BBC Future. https://www.bbc.com/future/article/20181218-whats-the-quickest-way-to-happiness-do-nothing.
29. Mauss, I.B., Tamir, M., Anderson, C.L., and Savino, N.S. (2011). Can seeking happiness make people unhappy? Paradoxical effects of valuing happiness. *Emotion*, *11*(4), 807. For a review of further research, see Gruber, J., Mauss, I.B., and Tamir, M. (2011). A dark side of happiness? How, when, and why happiness is not always good. *Perspectives on Psychological Science*, *6*(3), 222–33.
30. McGuirk, L., Kuppens, P., Kingston, R., and Bastian, B. (2018). Does a culture of happiness increase rumination over failure? *Emotion*, *18*(5), 755.
31. Ford, B.Q., Lam, P., John, O.P., and Mauss, I.B. (2018). The psychological health benefits of accepting negative emotions and thoughts: Laboratory, diary, and longitudinal evidence. *Journal of Personality and Social Psychology*, *115*(6), 1075. See also Shallcross, A.J., Troy, A.S., Boland, M., and Mauss, I.B. (2010). Let it be: Accepting negative emotional experiences predicts decreased negative affect and depressive symptoms. *Behaviour Research and Therapy*, *48*, 921–29.
32. Luong, G., Wrzus, C., Wagner, G.G., and Riediger, M. (2016). When bad moods may not be so bad: Valuing negative affect is associated with weakened affect-health links. *Emotion*, *16*(3), 387–401.
33. Tamir, M., and Bigman, Y.E. (2018). Expectations influence how emotions shape behavior. *Emotion*, *18*(1), 15. See also Tamir, M., and Ford, B.Q. (2012). When feeling bad is expected to be good: Emotion regulation and outcome expectancies in social conflicts. *Emotion*, *12*(4), 807.

34. Ford, B.Q., and Tamir, M. (2012). When getting angry is smart: Emotional preferences and emotional intelligence. *Emotion*, *12*(4), 685; Axt, J., and Oishi, S. (2016). When unfair treatment helps performance. *Motivation and Emotion*, *40*(2), 243–57.
35. Thakral, M., Von Korff, M., McCurry, S.M., Morin, C.M., and Vitiello, M.V. (2020). Changes in dysfunctional beliefs about sleep after cognitive behavioral therapy for insomnia: A systematic literature review and meta-analysis. *Sleep Medicine Reviews*, *49*, 101230. See also Courtauld, H., Notebaert, L., Milkins, B., Kyle, S.D., and Clarke, P.J. (2017). Individuals with clinically significant insomnia symptoms are characterised by a negative sleep-related expectancy bias: Results from a cognitive-experimental assessment. *Behaviour Research and Therapy*, *95*, 71–78.
36. Lichstein, K.L. (2017). Insomnia identity. *Behaviour Research and Therapy*, *97*, 230–41. See also Woosley, J.A., Lichstein, K.L., Taylor, D.J., Riedel, B.W., and Bush, A.J. (2016). Insomnia complaint versus sleep diary parameters: Predictions of suicidal ideation. *Suicide and Life-Threatening Behavior*, *46*(1), 88–95.
37. Draganich, C., and Erdal, K. (2014). Placebo sleep affects cognitive functioning. *Journal of Experimental Psychology: Learning, Memory, and Cognition*, *40*(3), 857; Gavriloff, D., Sheaves, B., Juss, A., Espie, C.A., Miller, C.B., and Kyle, S.D. (2018). Sham sleep feedback delivered via actigraphy biases daytime symptom reports in people with insomnia: Implications for insomnia disorder and wearable devices. *Journal of Sleep Research*, *27*(6), e12726. See also Rahman, S.A., Rood, D., Trent, N., Solet, J., Langer, E.J., and Lockley, S.W. (2020). Manipulating sleep duration perception changes cognitive performance—an exploratory analysis. *Journal of Psychosomatic Research*, *132*, 109992.
38. Personal communication with Kenneth Lichstein, University of Alabama, April 26, 2018.
39. https://www.cdc.gov/mmwr/volumes/68/wr/mm6849a5.htm.
40. Espie, C.A., Broomfield, N.M., MacMahon, K.M., Macphee, L.M., and Taylor, L.M. (2006). The attention-intention-effort pathway in the development of psychophysiologic insomnia: A theoretical review. *Sleep Medicine Reviews*, *10*(4), 215–45.
41. Thakral, M., Von Korff, M., McCurry, S.M., Morin, C.M., and Vitiello, M.V. (2020). Changes in dysfunctional beliefs about sleep after cognitive behavioral therapy for insomnia: A systematic literature review and meta-analysis. *Sleep Medicine Reviews*, *49*, 101230. See also Eidelman, P., Talbot, L., Ivers, H., Bélanger, L., Morin, C.M., and Harvey, A.G. (2016). Change in dysfunctional beliefs about sleep in behavior therapy, cognitive therapy, and cognitive-behavioral therapy for insomnia. *Behavior Therapy*, *47*(1), 102–15.
42. Selye, H. (1979). *The Stress of My Life: A Scientist's Memoirs*, 117. New York: Van Nostrand Reinhold. For more information about Selye's invention of the term "eustress," see Szabo, S., Tache, Y., and Somogyi, A. (2012). The legacy of Hans Selye and the origins of stress research: A retrospective 75 years after his landmark brief "letter" to the editor of *Nature*. *Stress*, *15*(5), 472–78.


## Chapter 8: Limitless Willpower


1. Lewis, M. (2012, September 11). Obama's way. *Vanity Fair*. https://www.vanityfair.com/news/2012/10/michael-lewis-profile-barack-obama.
2. Elkins, K. (2017, January 5). Billionaires Mark Zuckerberg and John Paul DeJoria use a simple wardrobe trick to boost productivity. CNBC. https://www.cnbc.com/2017/01/05/mark-zuckerberg-and-john-paul-dejorias-simple-wardrobe-trick.html.
3. De Vita, E. (2015, February 22). Creative thinking: Why a morning routine helps conserve your brainpower. *Financial Times*. https://www.ft.com/content/3d07fcea-b37b-11e4-9449-00144feab7de.

4. Baumeister, R.F., Bratslavsky, E., Muraven, M., and Tice, D.M. (1998). Ego depletion: Is the active self a limited resource? *Journal of Personality and Social Psychology*, *74*(5), 1252.
5. Ibid.
6. Inzlicht, M., Berkman, E., and Elkins-Brown, N. (2016). The neuroscience of "ego depletion." In M. Inzlicht, E. Berkman, N. Elkins-Brown (Eds.). *Social Neuroscience: Biological approaches to social psychology*, 101–23. London: Routledge.
7. Baumeister, R.F., Bratslavsky, E., Muraven, M., and Tice, D.M. (1998). Ego depletion: Is the active self a limited resource? *Journal of Personality and Social Psychology*, *74*(5), 1252.
8. Schmeichel, B.J., Vohs, K.D., and Baumeister, R.F. (2003). Intellectual performance and ego depletion: Role of the self in logical reasoning and other information processing. *Journal of Personality and Social Psychology*, *85*(1), 33; Schmeichel, B.J. (2007). Attention control, memory updating, and emotion regulation temporarily reduce the capacity for executive control. *Journal of Experimental Psychology: General*, *136*(2), 241.
9. Vohs, K.D., Baumeister, R.F., Schmeichel, B.J., Twenge, J.M., Nelson, N.M., and Tice, D.M. (2014). Making choices impairs subsequent self-control: A limited-resource account of decision making, self-regulation, and active initiative. *Motivation Science*, *1*(S), 19–42.
10. Vohs, K.D., and Faber, R.J. (2007). Spent resources: Self-regulatory resource availability affects impulse buying. *Journal of Consumer Research*, *33*(4), 537–47.
11. Baumeister, R.F. (2012). Self-control: The moral muscle. *The Psychologist*, *25*(2), 112–15. https://thepsychologist.bps.org.uk/volume-25/edition-2/self-control-%E2%80%93-moral-muscle.
12. Hofmann, W., Vohs, K.D., and Baumeister, R.F. (2012). What people desire, feel conflicted about, and try to resist in everyday life. *Psychological Science*, *23*(6), 582–88.
13. Baumeister, R.F., and Vohs, K.D. (2016). Strength model of self-regulation as limited resource: Assessment, controversies, update. *Advances in Experimental Social Psychology*, *54*, 67–127.
14. Parker, I. (2014, June 2). Inheritance. *New Yorker*. https://www.newyorker.com/magazine/2014/06/02/inheritance.
15. Sheppes, G., Catran, E., and Meiran, N. (2009). Reappraisal (but not distraction) is going to make you sweat: Physiological evidence for self-control effort. *International Journal of Psychophysiology*, *71*(2), 91–96; Wagstaff, C.R. (2014). Emotion regulation and sport performance. *Journal of Sport and Exercise Psychology*, *36*(4), 401–12.
16. See the following for a description of these PET scans, and Baumeister's own research in this area: Baumeister, R.F., and Vohs, K.D. (2016). Strength model of self-regulation as limited resource: Assessment, controversies, update. *Advances in Experimental Social Psychology*, *54*, 67–127.
17. Gailliot, M.T., Baumeister, R.F., DeWall, C.N., Maner, J.K., Plant, E.A., Tice, D.M., . . . and Schmeichel, B.J. (2007). Self-control relies on glucose as a limited energy source: Willpower is more than a metaphor. *Journal of Personality and Social Psychology*, *92*(2), 325.
18. Baumeister, R.F., and Vohs, K.D. (2016). Strength model of self-regulation as limited resource: Assessment, controversies, update. *Advances in Experimental Social Psychology*, *54*, 67–127.
19. For recent large-scale studies confirming the existence of ego depletion, see Dang, J., Liu, Y., Liu, X., and Mao, L. (2017). The ego could be depleted, providing initial exertion is depleting: A preregistered experiment of the ego depletion effect. *Social Psychology*, *48*(4), 242–45; Garrison, K. E., Finley, A.J., and Schmeichel, B. J. (2019). Ego depletion reduces attention control: Evidence from two high-powered preregistered experiments. *Personality and Social Psychology Bulletin*, *45*(5), 728–39; Dang, J., Barker, P., Baumert, A., Bentvel-

zen, M., Berkman, E., Buchholz, N., . . . and Zinkernagel, A. (2021). A multilab replication of the ego depletion effect. *Social Psychological and Personality Science, 12*(1), 14–24.

20. Martijn, C., Tenbült, P., Merckelbach, H., Dreezens, E., and de Vries, N.K. (2002). Getting a grip on ourselves: Challenging expectancies about loss of energy after self-control. *Social Cognition, 20*(6), 441–60. See also Clarkson, J.J., Hirt, E.R., Jia, L., and Alexander, M.B. (2010). When perception is more than reality: The effects of perceived versus actual resource depletion on self-regulatory behavior. *Journal of Personality and Social Psychology, 98*(1), 29. The following has a review of similar studies: Klinger, J.A., Scholer, A.A., Hui, C.M., and Molden, D.C. (2018). Effortful experiences of self-control foster lay theories that self-control is limited. *Journal of Experimental Social Psychology, 78*, 1–13.
21. Job, V., Dweck, C.S., and Walton, G.M. (2010). Ego depletion: Is it all in your head? Implicit theories about willpower affect self-regulation. *Psychological Science, 21*(11), 1686–93. See also Miller, E.M., Walton, G.M., Dweck, C.S., Job, V., Trzesniewski, K.H., and McClure, S.M. (2012). Theories of willpower affect sustained learning. *PLoS One, 7*(6), e38680; Chow, J.T., Hui, C.M., and Lau, S. (2015). A depleted mind feels inefficacious: Ego-depletion reduces self-efficacy to exert further self-control. *European Journal of Social Psychology, 45*(6), 754–68.
22. Bernecker, K., and Job, V. (2015). Beliefs about willpower moderate the effect of previous day demands on next day's expectations and effective goal striving. *Frontiers in Psychology, 6*, 1496.
23. See the longitudinal study in Job, V., Dweck, C.S., and Walton, G.M. (2010). Ego depletion: Is it all in your head? Implicit theories about willpower affect self-regulation. *Psychological Science, 21*(11), 1686–93. See also Job, V., Walton, G.M., Bernecker, K., and Dweck, C.S. (2015). Implicit theories about willpower predict self-regulation and grades in everyday life. *Journal of Personality and Social Psychology, 108*(4), 637; Bernecker, K., Herrmann, M., Brandstätter, V., and Job, V. (2017). Implicit theories about willpower predict subjective well-being. *Journal of Personality, 85*(2), 136–50.
24. Bernecker, K., and Job, V. (2015). Beliefs about willpower are related to therapy adherence and psychological adjustment in patients with type 2 diabetes. *Basic and Applied Social Psychology, 37*(3), 188–95. For a review of these findings, see also Job, V., Sieber, V., Rothermund, K., and Nikitin, J. (2018). Age differences in implicit theories about willpower: Why older people endorse a nonlimited theory. *Psychology and Aging, 33*(6), 940.
25. A full description of these experiments, along with hypotheses about the cultural origins of these mindsets and the effects on education, can be found in Savani, K., and Job, V. (2017). Reverse ego-depletion: Acts of self-control can improve subsequent performance in Indian cultural contexts. *Journal of Personality and Social Psychology, 113*(4), 589.
26. Scientific evidence supports the idea that trataka can improve concentration, potentially through the expectation effects that Job and Savani have described. See Raghavendra, B.R., and Singh, P. (2016). Immediate effect of yogic visual concentration on cognitive performance. *Journal of Traditional and Complementary Medicine, 6*(1), 34–36.
27. Descriptions of the conservation theory of ego depletion, and the evidence, can be found in Baumeister, R.F., and Vohs, K.D. (2016). Strength model of self-regulation as limited resource: Assessment, controversies, update. *Advances in Experimental Social Psychology, 54*, 67–127.
28. Job, V., Walton, G.M., Bernecker, K., and Dweck, C.S. (2013). Beliefs about willpower determine the impact of glucose on self-control. *Proceedings of the National Academy of Sciences, 110*(37), 14837–42.
29. Madzharov, A., Ye, N., Morrin, M., and Block, L. (2018). The impact of coffee-like scent on expectations and performance. *Journal of Environmental Psychology, 57*, 83–86; Denson,

T.F., Jacobson, M., Von Hippel, W., Kemp, R.I., and Mak, T. (2012). Caffeine expectancies but not caffeine reduce depletion-induced aggression. *Psychology of Addictive Behaviors*, *26*(1), 140; Cropsey, K.L., Schiavon, S., Hendricks, P.S., Froelich, M., Lentowicz, I., and Fargason, R. (2017). Mixed-amphetamine salts expectancies among college students: Is stimulant induced cognitive enhancement a placebo effect? *Drug and Alcohol Dependence*, *178*, 302–9.

30. Leach, S. (2019, May 9). How the hell has Danielle Steel managed to write 179 books? *Glamour.* https://www.glamour.com/story/danielle-steel-books-interview; Jordan, T. (2018, February 2). Danielle Steel: "I know an idea is right for me when it just clicks." *New York Times.* https://www.nytimes.com/2018/02/02/books/review/danielle-steel-fall-from-grace-best-seller.html.
31. Burkeman, O. (2019, May 31). Danielle Steel works 20 hours a day, but is that to be envied? *Guardian.* https://www.theguardian.com/money/oliver-burkeman-column/2019/may/31/danielle-steel-work-20-hour-day.
32. Konze, A.K., Rivkin, W., and Schmidt, K.H. (2019). Can faith move mountains? How implicit theories about willpower moderate the adverse effect of daily emotional dissonance on ego-depletion at work and its spillover to the home-domain. *European Journal of Work and Organizational Psychology*, *28*(2), 37–149. See also the following paper for an example of the ways that ego depletion can destroy our free time: Reinecke, L., Hartmann, T., and Eden, A. (2014). The guilty couch potato: The role of ego depletion in reducing recovery through media use. *Journal of Communication*, *64*(4), 569–89.
33. Bernecker, K., and and Job, V. (2020). Too exhausted to go to bed: Implicit theories about willpower and stress predict bedtime procrastination. *British Journal of Psychology*, *111*(1), 126–47.
34. See experiment 4 in Savani, K., and Job, V. (2017). Reverse ego-depletion: Acts of self-control can improve subsequent performance in Indian cultural contexts. *Journal of Personality and Social Psychology*, *113*(4), 589.
35. Sieber, V., Flückiger, L., Mata, J., Bernecker, K., and Job, V. (2019). Autonomous goal striving promotes a nonlimited theory about willpower. *Personality and Social Psychology Bulletin*, *45*(8), 1295–307.
36. Klinger, J.A., Scholer, A.A., Hui, C.M., and Molden, D.C. (2018). Effortful experiences of self-control foster lay theories that self-control is limited. *Journal of Experimental Social Psychology*, *78*, 1–13.
37. Haimovitz, K., Dweck, C.S., and Walton, G.M. (2020). Preschoolers find ways to resist temptation after learning that willpower can be energizing. *Developmental Science*, *23*(3), e12905.
38. On Williams's ritual: Serena Williams sings *Flashdance* theme to keep her calm on court. Sky News (2015, July 12). https://www.skysports.com/tennis/news/32498/9910795/serena-williams-sings-flashdance-theme-to-keep-her-calm-on-court. On Dr. Seuss and Beethoven: Weinstein, E. (2018, April 13). Ten superstitions of writers and artists. *Paris Review.* On Williams, Farrell, and Beyoncé: Brooks, A.W., Schroeder, J., Risen, J.L., Gino, F., Galinsky, A.D., Norton, M.I., and Schweitzer, M.E. (2016). Don't stop believing: Rituals improve performance by decreasing anxiety. *Organizational Behavior and Human Decision Processes*, *137*, 71–85. See also Hobson, N.M., Schroeder, J., Risen, J.L., Xygalatas, D., and Inzlicht, M. (2018). The psychology of rituals: An integrative review and process-based framework. *Personality and Social Psychology Review*, *22*(3), 260–84.
39. Lonsdale, C., and Tam, J.T. (2008). On the temporal and behavioural consistency of pre-performance routines: An intra-individual analysis of elite basketball players' free throw shooting accuracy. *Journal of Sports Sciences*, *26*(3), 259–66.

40. Damisch, L., Stoberock, B., and Mussweiler, T. (2010). Keep your fingers crossed! How superstition improves performance. *Psychological Science*, *21*(7), 1014–20.
41. Friese, M., Schweizer, L., Arnoux, A., Sutter, F., and Wänke, M. (2014). Personal prayer counteracts self-control depletion. *Consciousness and Cognition*, *29*, 90–95.
42. Rounding, K., Lee, A., Jacobson, J. A., & Ji, L. J. (2012). Religion replenishes self-control. *Psychological Science*, *23*(6), 635–42.
43. Brooks, A.W., Schroeder, J., Risen, J.L., Gino, F., Galinsky, A.D., Norton, M.I., and Schweitzer, M.E. (2016). Don't stop believing: Rituals improve performance by decreasing anxiety. *Organizational Behavior and Human Decision Processes*, *137*, 71–85.
44. Tian, A.D., Schroeder, J., Häubl, G., Risen, J.L., Norton, M.I., and Gino, F. (2018). Enacting rituals to improve self-control. *Journal of Personality and Social Psychology*, *114*(6), 851.

## Chapter 9: Untapped Genius

1. In the scientific literature, the school is known as Oak School, but an article in *Discover* magazine revealed the true location: Ellison, K. (2015, October 29). Being honest about the Pygmalion effect. *Discover*. https://www.discovermagazine.com/mind/being-honest-about-the-pygmalion-effect.
2. Rosenthal, R., and Jacobson, L. (1968). *Pygmalion in the Classroom: Teacher Expectation and Pupils' Intellectual Development*, 85–93. New York: Holt, Rinehart and Winston.
3. Rosenthal, R., and Jacobson, L. (1966). Teachers' expectancies: Determinants of pupils' IQ gains. *Psychological Reports*, *19*(1), 115–18.
4. See, for instance, Rudebeck, S.R., Bor, D., Ormond, A., O'Reilly, J.X., and Lee, A.C. (2012). A potential spatial working memory training task to improve both episodic memory and fluid intelligence. *PLoS One*, *7*(11), e50431.
5. Boot, W.R., Simons, D.J., Stothart, C., and Stutts, C. (2013). The pervasive problem with placebos in psychology: Why active control groups are not sufficient to rule out placebo effects. *Perspectives on Psychological Science*, *8*(4), 445–54.
6. Foroughi, C.K., Monfort, S.S., Paczynski, M., McKnight, P.E., and Greenwood, P.M. (2016). Placebo effects in cognitive training. *Proceedings of the National Academy of Sciences*, *113*(27), 7470–74.
7. See also Jaeggi, S.M., Buschkuehl, M., Shah, P., and Jonides, J. (2014). The role of individual differences in cognitive training and transfer. *Memory and Cognition*, *42*(3), 464–80; Miller, E.M., Walton, G.M., Dweck, C.S., Job, V., Trzesniewski, K.H., and McClure, S.M. (2012). Theories of willpower affect sustained learning. *PLoS One*, *7*(6), e38680.
8. Turi, Z., Bjørkedal, E., Gunkel, L., Antal, A., Paulus, W., and Mittner, M. (2018). Evidence for cognitive placebo and nocebo effects in healthy individuals. *Scientific Reports*, *8*(1), 1–14; Fassi, L., and Kadosh, R.C. (2020). Is it all in our head? When subjective beliefs about receiving an intervention are better predictors of experimental results than the intervention itself. *bioRxiv*. https://www.biorxiv.org/content/10.1101/2020.12.06.411850v1.abstract.
9. How drinking vodka makes you more creative. (2012, February 16) *The Week*. https://theweek.com/articles/478116/how-drinking-vodka-makes-more-creative.
10. Lipnicki, D.M., and Byrne, D.G. (2005). Thinking on your back: Solving anagrams faster when supine than when standing. *Cognitive Brain Research*, *24*(3), 719–22.
11. Lapp, W.M., Collins, R.L., and Izzo, C.V. (1994). On the enhancement of creativity by alcohol: Pharmacology or expectation? *American Journal of Psychology*, *107*(2), 173–206.

12. Rozenkrantz, L., Mayo, A.E., Ilan, T., Hart, Y., Noy, L., and Alon, U. (2017). Placebo can enhance creativity. *PLoS One*, *12*(9), e0182466. See also Weinberger, A.B., Iyer, H., and Green, A.E. (2016). Conscious augmentation of creative state enhances "real" creativity in open-ended analogical reasoning. *PLoS One*, e0150773.
13. Weger, U.W., and Loughnan, S. (2013). Rapid communication: Mobilizing unused resources: Using the placebo concept to enhance cognitive performance. *Quarterly Journal of Experimental Psychology*, *66*(1), 23–28.
14. Autin, F., and Croizet, J.C. (2012). Improving working memory efficiency by reframing metacognitive interpretation of task difficulty. *Journal of Experimental Psychology: General*, *141*(4), 610. See also Oyserman, D., Elmore, K., Novin, S., Fisher, O., and Smith, G.C. (2018). Guiding people to interpret their experienced difficulty as importance highlights their academic possibilities and improves their academic performance. *Frontiers in Psychology*, *9*, 781.
15. Rosenthal addresses some of the common criticisms in the following paper: Rosenthal, R. (1987). Pygmalion effects: Existence, magnitude, and social importance. *Educational Researcher*, *16*(9), 37–40. See also De Boer, H., Bosker, R.J., and van der Werf, M.P. (2010). Sustainability of teacher expectation bias effects on long-term student performance. *Journal of Educational Psychology*, *102*(1), 168. For a modern review, see Timmermans, A.C., Rubie-Davies, C.M., and Rjosk, C. (2018) Pygmalion's 50th anniversary: The state of the art in teacher expectation research. *Educational Research and Evaluation*, *24*(3–5), 91–98.
16. Szumski, G., and Karwowski, M. (2019). Exploring the Pygmalion effect: The role of teacher expectations, academic self-concept, and class context in students' math achievement. *Contemporary Educational Psychology*, *59*, 101787. See the following for a more critical review, which nevertheless finds that self-fulfilling prophecies are meaningful (and particularly high in the military): Jussim, L. (2017). Précis of social perception and social reality: Why accuracy dominates bias and self-fulfilling prophecy. *Behavioral and Brain Sciences*, *40*, e1.
17. Sorhagen, N.S. (2013). Early teacher expectations disproportionately affect poor children's high school performance. *Journal of Educational Psychology*, *105*(2), 465.
18. Eden, D., and Shani, A.B. (1982). Pygmalion goes to boot camp: Expectancy, leadership, and trainee performance. *Journal of Applied Psychology*, *67*(2), 194.
19. The effect size of the IDF study, and the average effect size across industries, can be found in the following paper: McNatt, D.B. (2000). Ancient Pygmalion joins contemporary management: A meta-analysis of the result. *Journal of Applied Psychology*, *85*(2), 314. For a further discussion of Pygmalion effects in the workplace, see Whiteley, P., Sy, T., and Johnson, S.K. (2012). Leaders' conceptions of followers: Implications for naturally occurring Pygmalion effects. *Leadership Quarterly*, *23*(5), 822–34; and Avolio, B.J., Reichard, R.J., Hannah, S.T., Walumbwa, F.O., and Chan, A. (2009). A meta-analytic review of leadership impact research: Experimental and quasi-experimental studies. *Leadership Quarterly*, *20*(5), 764–84.
20. Brophy, J.E., and Good, T.L. (1970). Teachers' communication of differential expectations for children's classroom performance: Some behavioral data. *Journal of Educational Psychology*, *61*(5), 365.
21. Rubie-Davies, C.M. (2007). Classroom interactions: Exploring the practices of high- and low-expectation teachers. *British Journal of Educational Psychology*, *77*(2), 289–306. For a comprehensive review, see Wang, S., Rubie-Davies, C.M., and Meissel, K. (2018). A systematic review of the teacher expectation literature over the past 30 years. *Educational Research and Evaluation*, *24*(3–5), 124–79.
22. Rosenthal, R., and Jacobson, L.F. (1968). Teacher expectations for the disadvantaged. *Scientific American*, *218*(4), 19–23.

23. As the following review explains, recent research shows that teacher expectations are stable over time: Timmermans, A.C., Rubie-Davies, C.M., and Rjosk, C. (2018). Pygmalion's 50th anniversary: The state of the art in teacher expectation research. *Educational Research and Evaluation*, *24*(3–5), 91–98.
24. Angelou, M. (2020). *I Know Why the Caged Bird Sings*, 83. London: Folio Society.
25. The teachers who changed Oprah's life. 1989. https://www.oprah.com/oprahshow/the-teachers-who-changed-oprahs-life/all.
26. Coughlan, S. (2016, March 8). Stephen Hawking remembers best teacher. BBC News. https://www.bbc.co.uk/news/education-35754759.
27. Talamas, S.N., Mavor, K.I., and Perrett, D.I. (2016). Blinded by beauty: Attractiveness bias and accurate perceptions of academic performance. *PLoS One*, *11*(2), e0148284. The authors make a direct connection to the expectation effect: "Perceptions of conscientiousness, intelligence and academic performance may play a vital role in the classroom environment and in the success of a child's education."
28. See, for example, Todorov, A., Mandisodza, A.N., Goren, A., and Hall, C.C. (2005). Inferences of competence from faces predict election outcomes. *Science*, *308*(5728), 1623–26; Moore, F.R., Filippou, D., and Perrett, D.I. (2011). Intelligence and attractiveness in the face: Beyond the attractiveness halo effect. *Journal of Evolutionary Psychology*, *9*(3), 205–17.
29. See Jæger, M.M. (2011). "A thing of beauty is a joy forever"? Returns to physical attractiveness over the life course. *Social Forces*, *89*(3), 983–1003; Frevert, T.K., and Walker, L.S. (2014). Physical attractiveness and social status. *Sociology Compass*, *8*(3), 313–23.
30. Clifford, M.M., and Walster, E. (1973). The effect of physical attractiveness on teacher expectations. *Sociology of Education*, 248–58; Bauldry, S., Shanahan, M.J., Russo, R., Roberts, B.W., and Damian, R. (2016). Attractiveness compensates for low status background in the prediction of educational attainment. *PLoS One*, *11*(6), e0155313.
31. Frieze, I.H., Olson, J.E., and Russell, J. (1991). Attractiveness and income for men and women in management. *Journal of Applied Social Psychology*, *21*(13), 1039–57. For a more in-depth discussion, see Toledano, E. (2013). May the best (looking) man win: The unconscious role of attractiveness in employment decisions. *Cornell HR Review*. http://digitalcommons.ilr.cornell.edu/chrr/48.
32. Mayew, W.J., Parsons, C.A., and Venkatachalam, M. (2013). Voice pitch and the labor market success of male chief executive officers. *Evolution and Human Behavior*, *34*(4), 243–48. Additional information, such as Skinner's earnings, comes from supplementary material attached to the paper, and an interview I conducted with William Mayew for the following video: Does the way you speak reveal how much you earn? (2018, June 5). BBC Worklife. https://www.bbc.com/worklife/article/20180605-does-the-way-you-speak-give-away-how-much-you-earn.
33. Wang, S., Rubie-Davies, C.M., and Meissel, K. (2018). A systematic review of the teacher expectation literature over the past 30 years. *Educational Research and Evaluation*, *24*(3–5), 124–79; Sorhagen, N.S. (2013). Early teacher expectations disproportionately affect poor children's high school performance. *Journal of Educational Psychology*, *105*(2), 465.
34. Jamil, F.M., Larsen, R.A., and Hamre, B.K. (2018). Exploring longitudinal changes in teacher expectancy effects on children's mathematics achievement. *Journal for Research in Mathematics Education*, *49*(1), 57–90.
35. Agirdag, O. (2018). The impact of school SES composition on science achievement and achievement growth: Mediating role of teachers' teachability culture. *Educational Research and Evaluation*, *24*(3–5), 264–76.
36. There has been a debate over the importance of stereotype threat, with some failed attempts to replicate the phenomenon. Proponents, however, argue that there have been method-

ological issues with some of those replications, and that the evidence for the existence of stereotype threat in many high-stakes situations is robust. Bolstering this argument, a recent meta-analysis confirmed that measures to reduce stereotype threat significantly boost performance among the people who would be at risk. For further information, see Nussbaum, D. (2018, February 1). The replicability issue and stereotype threat research. *Medium.* https://medium.com/@davenuss79/the-replicability-issue-and-stereotype-threat-research-a988d6f8b080; and Liu, S., Liu, P., Wang, M., and Zhang, B. (2020). Effectiveness of stereotype threat interventions: A meta-analytic review. *Journal of Applied Psychology.* https://doi.org/10.1037/apl0000770.

37. Quoted in: Ellison, K. (2015, October 29). Being honest about the Pygmalion effect. *Discover.* https://www.discovermagazine.com/mind/being-honest-about-the-pygmalion-effect.
38. Rubie-Davies, C.M., Peterson, E.R., Sibley, C.G., and Rosenthal, R. (2015). A teacher expectation intervention: Modelling the practices of high expectation teachers. *Contemporary Educational Psychology, 40,* 72–85. The data was re-analyzed in the following paper, which gives the 28 percent improvement quoted in this paragraph: Rubie-Davies, C.M., and Rosenthal, R. (2016). Intervening in teachers' expectations: A random effects meta-analytic approach to examining the effectiveness of an intervention. *Learning and Individual Differences, 50,* 83–92.
39. De Boer, H., Timmermans, A.C., and Van Der Werf, M.P. (2018). The effects of teacher expectation interventions on teachers' expectations and student achievement: Narrative review and meta-analysis. *Educational Research and Evaluation, 24*(3–5), 180–200.
40. John-Henderson, N.A., Rheinschmidt, M.L., and Mendoza-Denton, R. (2015). Cytokine responses and math performance: The role of stereotype threat and anxiety reappraisals. *Journal of Experimental Social Psychology, 56,* 203–6. Similar benefits can be seen for poorer students who might find examinations particularly stressful: Rozek, C.S., Ramirez, G., Fine, R.D., and Beilock, S.L. (2019). Reducing socioeconomic disparities in the STEM pipeline through student emotion regulation. *Proceedings of the National Academy of Sciences, 116*(5), 1553–58. See also Liu, S., Liu, P., Wang, M., and Zhang, B. (2020). Effectiveness of stereotype threat interventions: A meta-analytic review. *Journal of Applied Psychology, 106*(6), 921–49. doi: 10.1037/apl0000770.
41. The paper explicitly links it to research into expectation and stress. Brady, S.T., Reeves, S.L., Garcia, J., Purdie-Vaughns, V., Cook, J.E., Taborsky-Barba, S., . . . and Cohen, G.L. (2016). The psychology of the affirmed learner: Spontaneous self-affirmation in the face of stress. *Journal of Educational Psychology, 108*(3), 353.
42. Martens, A., Johns, M., Greenberg, J., and Schimel, J. (2006). Combating stereotype threat: The effect of self-affirmation on women's intellectual performance. *Journal of Experimental Social Psychology, 42*(2), 236–43.
43. Miyake, A., Kost-Smith, L.E., Finkelstein, N.D., Pollock, S.J., Cohen, G.L., and Ito, T.A. (2010). Reducing the gender achievement gap in college science: A classroom study of values affirmation. *Science, 330*(6008), 1234–37. Data on the gender gap taken from graph and supplementary material available here: www.sciencemag.org/cgi/content/full/330/6008/1234/DC1.
44. Hadden, I.R., Easterbrook, M.J., Nieuwenhuis, M., Fox, K.J., and Dolan, P. (2020). Self-affirmation reduces the socioeconomic attainment gap in schools in England. *British Journal of Educational Psychology, 90*(2), 517–36.
45. Cohen, G.L., Garcia, J., Apfel, N., and Master, A. (2006). Reducing the racial achievement gap: A social-psychological intervention. *Science, 313*(5791), 1307–10; Cohen, G.L., Garcia, J., Purdie-Vaughns, V., Apfel, N., and Brzustoski, P. (2009). Recursive processes in self-affirmation: Intervening to close the minority achievement gap. *Science, 324*(5925), 400–3.

46. Goyer, J.P., Garcia, J., Purdie-Vaughns, V., Binning, K.R., Cook, J.E., Reeves, S.L., . . . and Cohen, G.L. (2017). Self-affirmation facilitates minority middle schoolers' progress along college trajectories. *Proceedings of the National Academy of Sciences, 114*(29), 7594–99. See also Sherman, D.K., Hartson, K.A., Binning, K.R., Purdie-Vaughns, V., Garcia, J., Taborsky-Barba, S., . . . and Cohen, G.L. (2013). Deflecting the trajectory and changing the narrative: How self-affirmation affects academic performance and motivation under identity threat. *Journal of Personality and Social Psychology, 104*(4), 591. See the following for a summary of these studies on racial differences: Walton, G.M., and Wilson, T.D. (2018). Wise interventions: Psychological remedies for social and personal problems. *Psychological Review, 125*(5), 617.
47. For a meta-analysis of self-affirmation interventions, see Liu, S., Liu, P., Wang, M., and Zhang, B. (2020). Effectiveness of stereotype threat interventions: A meta-analytic review. *Journal of Applied Psychology*. For a description of the virtuous cycle, see Cohen, G.L., and Sherman, D.K. (2014). The psychology of change: Self-affirmation and social psychological intervention. *Annual Review of Psychology, 65*(1), 333–71.
48. Liu, S., Liu, P., Wang, M., and Zhang, B. (2020). Effectiveness of stereotype threat interventions: A meta-analytic review. *Journal of Applied Psychology*.

## Chapter 10: The Super-Agers

1. Gagliardi, S. (2018, February 6). Sanremo 2018. *Huffpost*. https://www.huffingtonpost.it/entry/sanremo-2018-paddy-jones-balla-a-83-anni-e-lascia-tutti-a-bocca-aperta-questanno-sanremo-lo-vince-lei-la-vecchia-che-balla-e-come-la-scimmia-di-gabbani_it_5cc1ef3ee4b0aa856c9ea862.
2. Yaqoob, J. (2014, April 12). Simon Cowell: Controversial salsa-dancing granny can win Britain's Got Talent—and she reminds me of mum. *Mirror*. https://www.mirror.co.uk/tv/tv-news/britains-talent-paddy-nico-simon-3406432.
3. This may sound controversial, but it's the conclusion of many papers, such as Stewart, T.L., Chipperfield, J.G., Perry, R.P., and Weiner, B. (2012). Attributing illness to "old age": Consequences of a self-directed stereotype for health and mortality. *Psychology and Health, 27*(8), 881–97.
4. The experiment is described in depth in Langer, E.J. (2009). *Counter Clockwise: Mindful Health and the Power of Possibility*. New York: Ballantine. Further details, including a discussion of future work, come from Pagnini, F., Cavalera, C., Volpato, E., Comazzi, B., Riboni, F.V., Valota, C., . . . and Langer, E. (2019). Ageing as a mindset: A study protocol to rejuvenate older adults with a counterclockwise psychological intervention. *BMJ Open, 9*(7), e030411.
5. Levy, B.R., Slade, M.D., Kunkel, S.R., and Kasl, S.V. (2002). Longevity increased by positive self-perceptions of aging. *Journal of Personality and Social Psychology, 83*(2), 261.
6. Levy, B.R., Zonderman, A.B., Slade, M.D., and Ferrucci, L. (2009). Age stereotypes held earlier in life predict cardiovascular events in later life. *Psychological Science, 20*(3), 296–98.
7. Levy, B.R., Ferrucci, L., Zonderman, A.B., Slade, M.D., Troncoso, J., and Resnick, S.M. (2016). A culture-brain link: Negative age stereotypes predict Alzheimer's disease biomarkers. *Psychology and Aging, 31*(1), 82.
8. Levy, B.R., Slade, M.D., Pietrzak, R.H., and Ferrucci, L. (2018). Positive age beliefs protect against dementia even among elders with high-risk gene. *PLoS One, 13*(2), e0191004.
9. Levy, B.R., Slade, M.D., Kunkel, S.R., and Kasl, S.V. (2002). Longevity increased by positive self-perceptions of aging. *Journal of Personality and Social Psychology, 83*(2), 261.

10. Kuper, H., and Marmot, M. (2003). Intimations of mortality: Perceived age of leaving middle age as a predictor of future health outcomes within the Whitehall II study. *Age and Ageing*, *32*(2), 178–84. There is also experimental evidence for a short-term effect here: people are affected by ageist TV ads, but only if they identify as being of the same generation as the actors: Westerhof, G.J., Harink, K., Van Selm, M., Strick, M., and Van Baaren, R. (2010). Filling a missing link: The influence of portrayals of older characters in television commercials on the memory performance of older adults. *Ageing and Society*, *30*(5), 897.
11. Stephan, Y., Sutin, A.R., and Terracciano, A. (2016). Feeling older and risk of hospitalization: Evidence from three longitudinal cohorts. *Health Psychology*, *35*(6), 634; Stephan, Y., Caudroit, J., Jaconelli, A., and Terracciano, A. (2014). Subjective age and cognitive functioning: A 10-year prospective study. *American Journal of Geriatric Psychiatry*, *22*(11), 1180–87.
12. Mock, S.E., and Eibach, R.P. (2011). Aging attitudes moderate the effect of subjective age on psychological well-being: Evidence from a 10-year longitudinal study. *Psychology and Aging*, *26*(4), 979. See the following papers for an elaboration of the link between subjective aging, psychological well-being, and physical health: Stephan, Y., Chalabaev, A., Kotter-Grühn, D., and Jaconelli, A. (2013). "Feeling younger, being stronger": An experimental study of subjective age and physical functioning among older adults. *Journals of Gerontology Series B: Psychological Sciences and Social Sciences*, *68*(1), 1–7; Westerhof, G.J., Miche, M., Brothers, A.F., Barrett, A.E., Diehl, M., Montepare, J.M., . . . and Wurm, S. (2014). The influence of subjective aging on health and longevity: A meta-analysis of longitudinal data. *Psychology and Aging*, *29*(4), 793; Wurm, S., and Westerhof, G.J. (2015). Longitudinal research on subjective aging, health, and longevity: Current evidence and new directions for research. *Annual Review of Gerontology and Geriatrics*, *35*(1), 145–65; Terracciano, A., Stephan, Y., Aschwanden, D., Lee, J.H., Sesker, A.A., Strickhouser, J.E., . . . and Sutin, A.R. (2021). Changes in subjective age during COVID-19. *Gerontologist*, *61*(1), 13–22.
13. Davies, C. (2010, January 24). Martin Amis in new row over "euthanasia booths." *Guardian*. https://www.theguardian.com/books/2010/jan/24/martin-amis-euthanasia-booths-alzheimers.
14. Martin Amis always had a fear and loathing of ageing. (2012, April 13). *Evening Standard*. https://www.standard.co.uk/news/martin-amis-always-had-a-fear-and-loathing-of-ageing-6791926.html. See also https://www.manchester.ac.uk/discover/news/writing-is-not-for-the-old-says-amis-yes-it-is-says-james.
15. Rosenbaum, R. (2012, August 31). Martin Amis contemplates evil. *Smithsonian*. https://www.smithsonianmag.com/arts-culture/martin-amis-contemplates-evil-17857756.
16. Higgins, C. (2009, January 24). Martin Amis on aging. *Guardian*. https://www.theguardian.com/books/2009/sep/29/martin-amis-the-pregnant-widow.
17. Levy, B. (2009). Stereotype embodiment: A psychosocial approach to aging. *Current Directions in Psychological Science*, *18*(6), 332–36.
18. Touron, D.R. (2015). Memory avoidance by older adults: When "old dogs" won't perform their "new tricks." *Current Directions in Psychological Science*, *24*(3), 170–76.
19. Robertson, D.A., King-Kallimanis, B.L., and Kenny, R.A. (2016). Negative perceptions of aging predict longitudinal decline in cognitive function. *Psychology and Aging*, *31*(1), 71; Jordano, M.L., and Touron, D.R. (2017). Stereotype threat as a trigger of mind-wandering in older adults. *Psychology and Aging*, *32*(3), 307.
20. Westerhof, G.J., Harink, K., Van Selm, M., Strick, M., and Van Baaren, R. (2010). Filling a missing link: The influence of portrayals of older characters in television commercials on the memory performance of older adults. *Ageing and Society*, *30*(5), 897.

21. Robertson, D.A., Savva, G.M., King-Kallimanis, B.L., and Kenny, R.A. (2015). Negative perceptions of aging and decline in walking speed: A self-fulfilling prophecy. *PLoS One*, *10*(4), e0123260.
22. Levy, B.R., and Slade, M.D. (2019). Positive views of aging reduce risk of developing later-life obesity. *Preventive Medicine Reports*, *13*, 196–98.
23. Stewart, T.L., Chipperfield, J.G., Perry, R.P., and Weiner, B. (2012). Attributing illness to "old age": Consequences of a self-directed stereotype for health and mortality. *Psychology and Health*, *27*(8), 881–97.
24. See, for instance, Levy, B.R., Ryall, A.L., Pilver, C.E., Sheridan, P.L., Wei, J.Y., and Hausdorff, J.M. (2008). Influence of African American elders' age stereotypes on their cardiovascular response to stress. *Anxiety, Stress, and Coping*, *21*(1), 85–93; Weiss, D. (2018). On the inevitability of aging: Essentialist beliefs moderate the impact of negative age stereotypes on older adults' memory performance and physiological reactivity. *Journals of Gerontology: Series B*, *73*(6), 925–33.
25. Levy, B.R., Moffat, S., Resnick, S.M., Slade, M.D., and Ferrucci, L. (2016). Buffer against cumulative stress: Positive age self-stereotypes predict lower cortisol across 30 years. *GeroPsych: The Journal of Gerontopsychology and Geriatric Psychiatry*, *29*(3), 141–46.
26. Levy, B.R., and Bavishi, A. (2018). Survival advantage mechanism: Inflammation as a mediator of positive self-perceptions of aging on longevity. *Journals of Gerontology: Series B*, *73*(3), 409–12.
27. https://www.newscientist.com/term/telomeres. See also Levitin, D. (2020) *The Changing Mind*, 325. London: Penguin Life.
28. Pietrzak, R.H., Zhu, Y., Slade, M.D., Qi, Q., Krystal, J.H., Southwick, S.M., and Levy, B.R. (2016). Negative age stereotypes' association with accelerated cellular aging: Evidence from two cohorts of older adults. *Journal of the American Geriatrics Society*, *64*(11), e228.
29. Tamman, A.J., Montalvo-Ortiz, J.L., Southwick, S.M., Krystal, J.H., Levy, B.R., and Pietrzak, R.H. (2019). Accelerated DNA methylation aging in US military veterans: Results from the National Health and Resilience in Veterans Study. *American Journal of Geriatric Psychiatry*, *27*(5), 528–32.
30. Levy, B.R., Slade, M.D., Pietrzak, R.H., and Ferrucci, L. (2018). Positive age beliefs protect against dementia even among elders with high-risk gene. *PLoS One*, *13*(2), e0191004.
31. Callaway, E. (2010, November 28). Telomerase reverses ageing process. *Nature*. doi: 10.1038/nature09603; Ledford, H. (2020). Reversal of biological clock restores vision in old mice. *Nature*, *88*(7837), 209.
32. Knechtle, B., Jastrzebski, Z., Rosemann, T., and Nikolaidis, P.T. (2019). Pacing during and physiological response after a 12-hour ultra-marathon in a 95-year-old male runner. *Frontiers in Physiology*, *9*, 1875.
33. Cited in this review paper: Lepers, R., and Stapley, P.J. (2016). Master athletes are extending the limits of human endurance. *Frontiers in Physiology*, *7*, 613.
34. Ibid.
35. Harvey-Wood, H. (2000, May 3). Obituary: Penelope Fitzgerald. *Guardian*. https://www.theguardian.com/news/2000/may/03/guardianobituaries.books.
36. Wood, J. (2014, November 17). Late bloom. *New Yorker*. https://www.newyorker.com/magazine/2014/11/24/late-bloom.
37. Sotheby's (2020). Getting to know Picasso ceramics. https://www.sothebys.com/en/articles/picasso-ceramics-7-things-you-need-to-know.
38. In pictures: Matisse's cut-outs. (2013, October 7). BBC News. https://www.bbc.co.uk/news/in-pictures-24402817.

39. Weiss, D. (2018). On the inevitability of aging: Essentialist beliefs moderate the impact of negative age stereotypes on older adults' memory performance and physiological reactivity. *Journals of Gerontology: Series B*, *73*(6), 925–33.
40. Shimizu, A. (2019, April 5). For Hiromu Inada, an 86-year-old ironman triathlete, age really is just a number. *Japan Times*: https://www.japantimes.co.jp/life/2019/04/05/lifestyle/hiromu-inada-86-year-old-ironman-triathlete-age-really-just-number.
41. (UK) Office for National Statistics (2018). Living longer: How our population is changing and why it matters. https://www.ons.gov.uk/releases/livinglongerhowourpopulationischangingandwhyitmatters.
42. https://www.who.int/news-room/fact-sheets/detail/dementia.
43. Kaeberlein, M. (2018). How healthy is the healthspan concept? *GeroScience*, *40*(4), 361–64.
44. Levy, B.R., Pilver, C., Chung, P.H., and Slade, M.D. (2014). Subliminal strengthening: Improving older individuals' physical function over time with an implicit-age-stereotype intervention. *Psychological Science*, *25*(12), 2127–35.
45. Robertson, D.A., King-Kallimanis, B.L., and Kenny, R.A. (2016). Negative perceptions of aging predict longitudinal decline in cognitive function. *Psychology and Aging*, *31*(1), 71–81.
46. Sarkisian, C.A., Prohaska, T.R., Davis, C., and Weiner, B. (2007). Pilot test of an attribution retraining intervention to raise walking levels in sedentary older adults. *Journal of the American Geriatrics Society*, *55*(11), 1842–46.
47. See, for instance, Stephan, Y., Chalabaev, A., Kotter-Grühn, D., and Jaconelli, A. (2013). "Feeling younger, being stronger": An experimental study of subjective age and physical functioning among older adults. *Journals of Gerontology Series B: Psychological Sciences and Social Sciences*, *68*(1), 1–7; Brothers, A., and Diehl, M. (2017). Feasibility and efficacy of the AgingPlus Program: Changing views on aging to increase physical activity. *Journal of Aging and Physical Activity*, *25*(3), 402–11; Nehrkorn-Bailey, A., Forsyth, G., Braun, B., Burke, K., and Diehl, M. (2020). Improving hand-grip strength and blood pressure in adults: Results from an AgingPLUS pilot study. *Innovation in Aging*, *4* (Suppl 1), 587; Wolff, J.K., Warner, L.M., Ziegelmann, J.P., and Wurm, S. (2014). What do targeting positive views on ageing add to a physical activity intervention in older adults? Results from a randomised controlled trial. *Psychology and Health*, *29*(8), 915–32; Beyer, A.K., Wolff, J.K., Freiberger, E., and Wurm, S. (2019). Are self-perceptions of ageing modifiable? Examination of an exercise programme with vs. without a self-perceptions of ageing-intervention for older adults. *Psychology and Health*, *34*(6), 661–76.
48. I have written about this research previously: Robson, D. (2017, August 28). The amazing fertility of the older mind. BBC Future. http://www.bbc.com/future/story/20170828-the-amazing-fertility-of-the-older-mind.
49. https://www.tuttitalia.it/sardegna/73-nuoro/statistiche/popolazione-andamento-demografico.
50. Kirchgaessner, S. (2016, August 12). Ethical questions raised in search for Sardinian centenarians' secrets. *Guardian*. https://www.theguardian.com/world/2016/aug/12/ethical-questions-raised-in-search-for-sardinian-centenarians-secrets; https://www.bluezones.com/exploration/sardinia-italy.
51. Ruby, J.G., Wright, K.M., Rand, K.A., Kermany, A., Noto, K., Curtis, D., . . . and Ball, C. (2018). Estimates of the heritability of human longevity are substantially inflated due to assortative mating. *Genetics*, *210*(3), 1109–24.
52. My short documentary on this subject can be found at https://www.bbc.com/reel/playlist/elixir-of-life?vpid=p08blgc4.

53. North, M.S., and Fiske, S.T. (2015). Modern attitudes toward older adults in the aging world: A cross-cultural meta-analysis. *Psychological Bulletin*, *141*(5), 993.
54. Levy, B.R. (2017). Age-stereotype paradox: Opportunity for social change. *Gerontologist*, *57* (Suppl 2), S118–S126.

## Epilogue

1. Anzilotti, E. (2017, March 7). This hospital bridges traditional medicine with Hmong spirituality—and gets results. *Fast Company*. https://www.fastcompany.com/3068680/this-hospital-bridges-traditional-medicine-with-hmong-spirtuality-and-gets-results.
2. Colucci-D'Amato, L., Bonavita, V., and Di Porzio, U. (2006). The end of the central dogma of neurobiology: Stem cells and neurogenesis in adult CNS. *Neurological Sciences*, *27*(4), 266–70.
3. Schroder, H.S., Kneeland, E.T., Silverman, A.L., Beard, C., and Björgvinsson, T. (2019). Beliefs about the malleability of anxiety and general emotions and their relation to treatment outcomes in acute psychiatric treatment. *Cognitive Therapy and Research*, *43*(2), 312–23.
4. Burnette, J.L. (2010). Implicit theories of body weight: Entity beliefs can weigh you down. *Personality and Social Psychology Bulletin*, *36*(3), 410–22; Burnette, J.L., and Finkel, E.J. (2012). Buffering against weight gain following dieting setbacks: An implicit theory intervention. *Journal of Experimental Social Psychology*, *48*(3), 721–25; Burnette, J.L., Knouse, L.E., Vavra, D.T., O'Boyle, E., and Brooks, M.A. (2020). Growth mindsets and psychological distress: A meta-analysis. *Clinical Psychology Review*, *77*, 101816.
5. See the supplemental material to Yeager, D.S., Johnson, R., Spitzer, B.J., Trzesniewski, K.H., Powers, J., and Dweck, C.S. (2014). The far-reaching effects of believing people can change: Implicit theories of personality shape stress, health, and achievement during adolescence. *Journal of Personality and Social Psychology*, *106*(6), 867.
6. Kross, E., and Ayduk, O. (2017). Self-distancing: Theory, research, and current directions. *Advances in Experimental Social Psychology*, *55*, 81–136.
7. Streamer, L., Seery, M.D., Kondrak, C.L., Lamarche, V.M., and Saltsman, T.L. (2017). Not I, but she: The beneficial effects of self-distancing on challenge/threat cardiovascular responses. *Journal of Experimental Social Psychology*, *70*, 235–41.
8. Diedrich, A., Hofmann, S.G., Cuijpers, P., and Berking, M. (2016). Self-compassion enhances the efficacy of explicit cognitive reappraisal as an emotion regulation strategy in individuals with major depressive disorder. *Behaviour Research and Therapy*, *82*, 1–10.

# ILLUSTRATION CREDITS

[page 14] (Illusion) From McCrone, J. (1991). *The ape that spoke: Language and the evolution of the human mind.* New York: William Morrow & Company.

[page 15] (Unresolved Picture) Courtesy of Nava Rubin. From: Ludmer, R., Dudai, Y., and Rubin, N. (2011). Uncovering camouflage: Amygdala activation predicts long-term memory of induced perceptual insight. *Neuron*, 69(5), 1002–14.

[page 16] (Duck/Rabbit) From *Fliegende Blätter*, October 23, 1892.

[page 33] (Resolved Picture) Courtesy of Nava Rubin. From: Ludmer, R., Dudai, Y., and Rubin, N. (2011). Uncovering camouflage: Amygdala activation predicts long-term memory of induced perceptual insight. *Neuron*, 69(5), 1002–14.

[page 123] (Change in Arm Strength) Source: Yao, W. X., Ranganathan, V. K., Allexandre, D., Siemionow, V., & Yue, G. H. (2013). Kinesthetic imagery training of forceful muscle contractions increases brain signal and muscle strength. *Frontiers in Human Neuroscience*, *7*, 561.

[page 136] (Hunger After "Tasty" and "Healthy" Chocolate Bars) Source: Finkelstein, S. R., & Fishbach, A. (2010). When healthy food makes you hungry. *Journal of Consumer Research*, *37*(3), 357–67.

[page 223] (Self-Affirmation Reduces Gender Differences in Spatial Reasoning) Source: Martens, A., Johns, M., Greenberg, J., and Schimel, J. (2006). Combating stereotype threat: The effect of self-affirmation on women's intellectual performance. *Journal of Experimental Social Psychology*, *42*(2), 236–43.

[page 224] (Self-Affirmation Reduces Gender Differences in Physics Performance) Source: Miyake, A., Kost-Smith, L. E., Finkelstein, N. D., Pollock, S. J., Cohen, G. L., and Ito, T. A. (2010). Reducing the gender achievement gap in college science: A classroom study of values affirmation. *Science*, 330(6008), 1234–37.

[page 234] (The Effects of Age Beliefs on Dementia Incidence) Source: Levy, B. R., Slade, M. D., Pietrzak, R. H., & Ferrucci, L. (2018). Positive age beliefs protect against dementia even among elders with high-risk gene. *PLoS One*, 13(2), e0191004.

## ACKNOWLEDGMENTS

*The Expectation Effect* has grown from the generosity of many people. Thanks first to my agent, Carrie Plitt, for her enthusiasm at my initial idea, her astute and tactful feedback through its many iterations, and for her passion throughout its development. I'm also indebted to the rest of the team at Felicity Bryan Associates, and to Zoë Pagnamenta in New York for finding a home for the book in the United States.

I'm enormously grateful to my two editors, Simon Thorogood at Canongate and Conor Mintzer at Henry Holt, for their wisdom and kindness—it's been a joy to work with you both. Thanks also to my copyeditors, Debs Warner and Helen Maggie Carr, for saving me from numerous infelicities—and to the production, marketing, publicity, and sales teams at Canongate and at Holt.

I will be eternally indebted to the scientists whose work I have cited. Thanks especially to all the researchers who took the time to talk me through their work on mindset and expectation. In alphabetical order: Moshe Bar, Andy Clark, Luana Colloca, Alia Crum, Grace Giles, Suzanne Higgs, Jeremy Jamieson, Veronika Job, Beat Knechtle, Johannes Laferton, Kari Leibowitz, Becca Levy, Iris Mauss, Timothy Noakes, Keith Petrie, Christine Rubie-Davies, Anil Seth, and Jon Stone. Thanks also to Paddy Jones for sharing her life story with me.

My initial conception of *The Expectation Effect* came from an article commissioned by Kate Douglas at *New Scientist*. Thank you for accepting my pitch, shaping the story, and setting the ball rolling. Richard Fisher provided early feedback on some particularly troublesome chapters—your comments helped me to see the wood for the trees. And my regular catch-ups with Melissa Hogenboom offered the perfect pep talk when I was feeling frustrated or demotivated, and made the writing process so much less lonely.

Thanks to my friends and colleagues, including Sally Adee, Lindsay Baker, Amy Charles, Eileen and Peter Davies, Kerry Daynes, Stephen Dowling, Natasha and Sam Fenwick, Philippa Fogarty, Simon Frantz, Alison George, Zaria Gorvett, Richard Gray, Christian Jarrett, Rebecca Laurence, Fiona Macdonald, Damiano Mirigliano, Will Park, Emma and Sam Partington, Jo Perry, Mithu Storoni, Neil and Lauren Sullivan, Ian Tucker, Meredith Turits, Gaia Vince, James Wallman, Richard Webb, and Clare Wilson.

I owe more than I can describe to my parents, Margaret and Albert. Thanks most of all to Robert Davies for your support in every step of this journey, and every other part of my life. I could not have written this book without you.

# INDEX

## ABOUT THE AUTHOR

David Robson is an award-winning science writer based in the United Kingdom. A graduate of Cambridge University, he previously worked as an editor at *New Scientist* and a senior journalist at the BBC. His writing has appeared in the *Guardian*, the *Atlantic*, *Men's Health*, the *Psychologist*, the *Washington Post*, and many other publications. His first book, *The Intelligence Trap*, was published in 2019 and has been translated into fifteen languages.